GOSS'S ROOFING READY RECKONER

METRIC CUTTING AND SIZING TABLES FOR TIMBER ROOF MEMBERS

Fourth edition
Revised by
C. N. Mindham

Blackwell
Publishing

© 1948, 1987, 2001, 2008 the estate of Ralph Goss, Blackwell Publishing and Chris N. Mindham

Blackwell Publishing editorial offices:
Blackwell Publishing Ltd, 9600 Garsington Road, Oxford OX4 2DQ, UK
  Tel: +44 (0)1865 776868
Blackwell Publishing Inc., 350 Main Street, Malden, MA 02148-5020, USA
  Tel: +1 781 388 8250
Blackwell Publishing Asia Pty Ltd, 550 Swanston Street, Carlton, Victoria 3053, Australia
  Tel: +61 (0)3 8359 1011

The right of the Author to be identified as the Author of this Work has been asserted in accordance with the Copyright, Designs and Patents Act 1988.

All rights reserved. No part of this publication may be reproduced, stored in a retrieval system, or transmitted, in any form or by any means, electronic, mechanical, photocopying, recording or otherwise, except as permitted by the UK Copyright, Designs and Patents Act 1988, without the prior permission of the publisher.

Designations used by companies to distinguish their products are often claimed as trademarks. All brand names and product names used in this book are trade names, service marks, trademarks or registered trademarks of their respective owners. The Publisher is not associated with any product or vendor mentioned in this book.

This publication is designed to provide accurate and authoritative information in regard to the subject matter covered. It is sold on the understanding that the Publisher is not engaged in rendering professional services. If professional advice or other expert assistance is required, the services of a competent professional should be sought.

First published 2008 by Blackwell Publishing Ltd
Second edition published 1987
Third edition published 2001
First edition published in 1948

2 2009

ISBN: 978-1-4051-5921-0

Library of Congress Cataloging-in-Publication Data

Goss, Ralph.
  Goss's roofing ready reckoner : metric cutting and sizing tables for timber roof members. – 4th ed. / Chris N. Mindham.
    p. cm.
  Rev. ed. of: Roofing ready reckoner / Ralph Goss. 2001.
  Includes bibliographical references and index.
  ISBN 978-1-4051-5921-0 (pbk. : alk. paper)
  1. Roofs–Handbooks, manuals, etc.
2. Carpentry–Mathematics–Handbooks, manuals, etc.
3. Roofing–Handbooks, manuals, etc.
4. Engineering mathematics–Formulae–Handbooks, manuals, etc.
I. Mindham, C. N. (Chris N.) II. Goss, Ralph. Roofing ready reckoner. III. Title.

TH2401.G67 2007
694′.2–dc22

2007008607

A catalogue record for this title is available from the British Library

Set in 9.5/11pt Universe
by Aptara Inc., New Delhi, India
Printed and bound in Singapore
by Fabulous Printers Pte Ltd

The publisher's policy is to use permanent paper from mills that operate a sustainable forestry policy, and which has been manufactured from pulp processed using acid-free and elementary chlorine-free practices. Furthermore, the publisher ensures that the text paper and cover board used have met acceptable environmental accreditation standards.

For further information on Blackwell Publishing, visit our website:
www.blackwellpublishing.com

CONTENTS

1	Introduction	1
2	Roofing terminology	2
3	Calculating the size of timber members	8
	Strength and section size calculation	9
	How do we calculate the loading on the roof?	15
	Timber member sizing design: example	15
4	Calculating the length and cutting angles of timber members: data tables 5°–75°	21
	The pitch	24
	Using the tables to cut a common rafter	24
	Hip jack rafters	29
	Hip rafters	31
	Valley jack rafters	33
	The ridge	33
	Purlins	35
	Metric calculation tables	36

Contents

5 Wall plates – strapping and gable strapping — 110

6 Wind bracing and openings for dormers and roof windows — 112
   Openings for dormers and roof windows — 114

7 Roofing metalwork and fixings — 120
   Nails, bolts and screws — 122

8 Roof coverings – felt, battens and tiles — 124
   Underlay — 124
   Battens — 125
   Insulation and ventilation — 129
   Choosing the roof covering — 139

9 Roof coverings – building detail drawings — 151
   Natural slates — 151
   Concrete interlocking tiles — 154
   Plain and peg tiles — 158
   Asphalt shingles — 164
   Metal tiles — 167

10 Tools and equipment — 170
- Obtaining information from the drawing — 170
- To cut the roof — 170
- Setting up the roof structure — 171
- Roof coverings — 171

11 Health & safety considerations — 173
- Access to the roof — 173
- Basic principles — 173
- Restoration and renovation of existing roof structures — 174
- Newly constructed roofs — 175
- The roof covering — 175
- Conclusion — 176

Bibliography — 177

Index — 183

1 INTRODUCTION

The aim of this book, when first published in 1948, was to provide quick reference tables for the length and angles of cut for timber members in a traditional cut roof construction. Today, when many houses use trussed rafters for their roof construction, there is still a need for some parts of those roofs to be built using traditional methods, especially with the ever increasing use of attic roof structures. Renovation of older roofs, extensions and conversions all require knowledge of roofing from wall plate to ridge and the correct detailing of the roof covering materials themselves.

This book assumes that a basic architectural design of the roof to be constructed is already completed, i.e. the span, pitch, length and any additional supporting walls. Guidance is given on how to calculate the size of individual roof member timbers, the cutting length, the angles and the compound cuts. The book now also includes all aspects to be considered when choosing the roof covering, including the suitability of the tiles or slates for the pitch and exposure of the roof concerned, the choice of a 'warm' or 'cold' roof, the considerations to be given to the correct insulation, and the possibilities and avoidance of condensation within the roof space by dealing correctly with ventilation.

Finally, Health and Safety matters are addressed, including the 'Working at Heights Regulations', loading the roof structure with the roof coverings, lifting components, and the correct use of preservative treated timber.

2 ROOFING TERMINOLOGY

Wall plate The 'foundation' of the roof, usually 50 × 100 mm wide, must be bedded solid, level and straight on the top of the wall, or nailed to the timber framed panel and strapped in place to prevent movement from the structure.

Purlin The member carrying part load of the long common rafters, traditionally placed at right angles to the rafter but now more commonly fixed vertically.

Pitch The angle made by the slope of the roof with the horizontal. This may be stated in degrees on the drawing, or may have to be measured by protractor from the drawing, or may have to be calculated by measurement if the new work is to match an existing roof.

Ridge The timber at the top of the roof where the rafters meet, giving a longitudinal tie to the roof structure, commonly 38 mm thick, and of a depth equal to the top cut on the rafter plus approximately 38 mm. This depth will depend upon the pitch of the roof and the tile batten thickness.

Common rafter The timber running from the ridge, down over the purlin if fitted, over the wall plate, and to the back of the fascia.

Jack rafter The timber running from the hip rafter down over the purlin if fitted, over the wall plate, and to the back of the fascia.

Roofing terminology

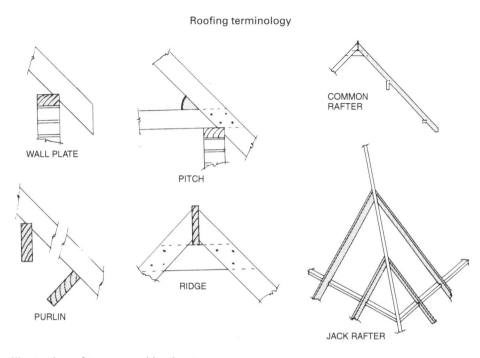

Figure 2.1 Illustration of terms used in chapter.

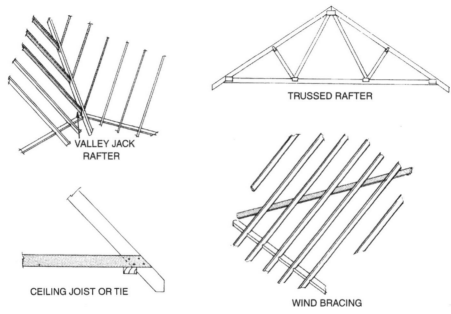

Figure 2.2 Illustration of terms used in chapter.

Valley jack rafter   The timber running from the ridge, down over the purlin, down to the valley board or rafter.

Trussed rafter   A prefabricated framework incorporating rafter, ceiling joist (or tie), and strengthening webs forming a fully triangulated structural element.

Ceiling joist or tie   Timber supporting the ceiling of the building, but often importantly 'tying' the feet of the common and jack rafters together thus triangulating and stabilising the roof.

Wind bracing   Usually 25 mm × 100 mm timber nailed to the underside of rafters and trussed rafters running at approximately 45° to them, to triangulate and stabilise the roof in its vertical plain.

Goss's roofing ready reckoner

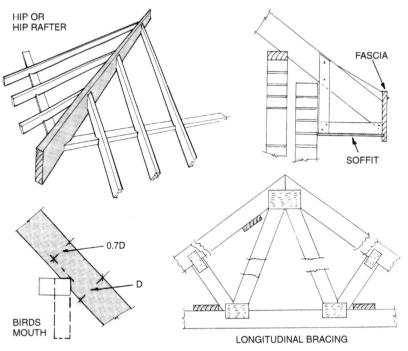

Figure 2.3 Illustration of terms used in chapter.

Longitudinal bracing   Usually 25 mm × 100 mm timber nailed to the underside of rafters and trussed rafters both at the ridge position on a trussed rafter roof, and at ceiling joist level on all roofs, to maintain accurate spacing and stiffening of the members to which it is fixed.

Hip or hip rafter   This is a substantial timber member running from the corner of the roof at wall plate level to the end of the ridge. In some designs the hip may stop lower down the roof, producing a small gable at high level.

Birds mouth   The cut in rafters at the fixing point to the wall plate and/or the purlin (where purlins are fixed vertically); this should leave at least 0.7 × the depth of the rafter to give the strength necessary for the rafter to continue to provide an over hang to the roof. If a common rafter is fitted as part of a trussed rafter roofing system, then the 0.7 × the depth of the rafter must be the same as the depth of the rafter on the trussed rafter component.

Fascia   Board fixed to the rafter feet, supporting both gutter and soffit.

Soffit   Timber board or sheet material used to close off the over hang between the back of the fascia and the wall. This soffit may have a roof ventilation system built into it.

3 CALCULATING THE SIZE OF TIMBER MEMBERS

Knowing the overall dimensions of the roof, i.e. span over the wall plates, pitch of the rafters, length between the gables or hips, any internal supporting walls and the specification of the roof covering, the following data will help to design the size and strength specification of the individual roof members themselves.

Design to satisfy the building regulations can be calculated from the TRADA document *Span Tables for Solid Timber Members in Floors, Ceilings and Roofs (Excluding Trussed Rafter Roofs) For Dwellings*. See Bibliography for full details of this publication. Some span tables from this publication are given on the following pages, covering a limited range of pitch and loadings for rafters, purlins, ceiling joist and binders (not floor joists). While these tables can be used for limited calculations, if the reader is likely to have a need to design timber members frequently, it is strongly recommended that the TRADA publication is available; this publication contains complete data, including all references to British Standards and codes of practice related to timber roof structures.

A brief word on timber size design considerations. The loads to be supported by the roof structure are made up of a number of elements:

(a) The roof covering: tiles, slates, etc.
(b) The self weight of the structure: timber, felt, battens, insulation, and ceiling if an attic structure, plus water tanks as necessary

(c) Snow load
(d) Wind load

Referring to the above:

(a) is a statistic relating to the roof covering type and can be obtained from the product manufacturer.
(b) is calculated from the known weight of individual members.
(c) is a variable with the roof shape and pitch and height above ground and sea level and is also a variable depending on the location of the building geographically within the UK.
(d) Similarly, this is a variable based on geographically variable information and is also affected by exposure, i.e. the altitude of the building and its proximity to coastal or other areas of high wind exposure.

For the above reasons, any calculation for specifying the roof member size must take all of the above into consideration. Thus, most standard timber sizing design data has geographical limitations to its use. The tables reproduced in this book from the TRADA publication cover buildings situated not exceeding 100 m above sea level for snow loading, and are confined to England and Wales. Refer to the TRADA publication for full details, but Figure 3.1 and Table 3.5 illustrate the basic information stated above.

STRENGTH AND SECTION SIZE CALCULATION
The load bearing capacity of a timber member is a function not only of its cross section, but also of its strength class. Readily available timbers are classified from strength class C16 and C24; these include a range of European, UK, Canadian and USA produced timbers. Whilst section size savings can be achieved using the higher grade timbers, there is a cost premium to pay, and on small scale projects, the economy of timber section is not great; more detail on this later. The C16 timber, being of less strength than the C24, results in a larger section, and in some cases the greater width of the timber

can be of benefit to the non-professional, giving a greater width to which to fix battens, etc. It is for this reason that tables for C16 timber only have been published in this book, (Tables 3.1–3.5) but a full range of tables for both C16 and C24 strength class timber can be found in the TRADA publication.

Table 3.1 Permissible clear spans for ceiling joists. **Imposed load not exceeding 0.25 kN/m^2; strength class C16; service class 1 or 2.**

Size of joist		Dead load (kN/m^2) excluding self-weight of joist					
		Not more than 0.25			More than 0.25 but not more than 0.5		
		Spacing of joists (mm)					
Breadth (mm)	Depth (mm)	400	450	600	400	450	600
		Maximum clear span (m)					
38	72	1.15	1.14	1.11	1.11	1.10	1.06
38	97	1.74	1.72	1.67	1.67	1.65	1.58
38	120	2.33	2.29	2.21	2.21	2.17	2.08
38	145	2.98	2.94	2.82	2.82	2.76	2.62
38	170	3.66	3.60	3.43	3.43	3.36	3.18
38	195	4.34	4.26	4.05	4.05	3.97	3.74
38	220	5.03	4.93	4.68	4.68	4.57	4.30
47	72	1.27	1.26	1.23	1.23	1.22	1.18
47	97	1.93	1.90	1.84	1.84	1.81	1.74

Table 3.1 Continued.

Size of joist		Dead load (kN/m^2) excluding self-weight of joist					
		Not more than 0.25			More than 0.25 but not more than 0.5		
		Spacing of joists (mm)					
Breadth (mm)	Depth (mm)	400	450	600	400	450	600
		Maximum clear span (m)					
47	120	2.56	2.52	2.43	2.43	2.38	2.27
47	145	3.27	3.22	3.08	3.08	3.02	2.87
47	170	4.00	3.93	3.74	3.74	3.67	3.46
47	195	4.73	4.64	4.41	4.41	4.31	4.07
47	220	5.47	5.36	5.08	5.08	4.96	4.67
ALS/CLS							
38	89	1.55	1.53	1.49	1.49	1.47	1.41
38	140	2.85	2.81	2.69	2.69	2.64	2.51
38	184	4.04	3.97	3.78	3.78	3.70	3.49

Reproduced from *Span Tables for Solid Timber Members in Floors, Ceilings and Roofs (Excluding Trussed Rafter Roofs) for Dwellings* by permission of TRADA Technology Ltd. (Table 8, p 19)

Table 3.2 Permissible clear spans for ceiling binders. **Imposed load 0.25 kN/m^2; strength class C16; service class 1 or 2.**

Size of binder		Dead load (kN/m^2) excluding self-weight of binder and ceiling joist										
		Not more than 0.25						More than 0.25 but not more than 0.5					
		Spacing of binders (mm)											
Breadth (mm)	Depth (mm)	1200	1500	1800	2100	2400	2700	1200	1500	1800	2100	2400	2700
		Maximum clear span (m)											
47	150	2.20	2.08	1.99	1.91	1.84		2.01	1.89	1.80			
47	175	2.63	2.49	2.37	2.27	2.19	2.11	2.39	2.25	2.14	2.04	1.96	1.89
63	125	2.00	1.90	1.82				1.84					
63	150	2.48	2.35	2.24	2.15	2.07	2.00	2.26	2.13	2.03	1.94	1.86	
63	175	2.95	2.79	2.66	2.55	2.46	2.37	2.69	2.53	2.40	2.29	2.20	2.12
63	200	3.43	3.24	3.08	2.95	2.84	2.74	3.11	2.92	2.77	2.65	2.54	2.45
63	225	3.91	3.68	3.50	3.35	3.22	3.11	3.54	3.32	3.14	3.00	2.88	2.77
75	125	2.15	2.04	1.95	1.88	1.81		1.97	1.86				
75	150	2.65	2.51	2.40	2.30	2.22	2.15	2.42	2.28	2.17	2.07	1.99	1.92
75	175	3.16	2.98	2.84	2.72	2.62	2.54	2.87	2.70	2.56	2.45	2.35	2.27
75	200	3.66	3.45	3.29	3.15	3.03	2.93	3.32	3.12	2.96	2.83	2.71	2.62
75	225	4.16	3.93	3.73	3.57	3.44	3.32	3.77	3.54	3.36	3.20	3.07	2.96

Reproduced from *Span Tables for Solid Timber Members in Floors, Ceilings and Roofs (Excluding Trussed Rafter Roofs) for Dwellings* by permission of TRADA Technology Ltd. (Table 10, p 21)

Calculating size of timber

Table 3.3 Permissible clear spans for purlins supporting rafters. **Imposed load 0.75 kN/m^2; slope of roof 22.5° or more but less than 30.0°; strength class C16; service class 1 or 2.**

Size of purlin		Dead load (kN/m^2) excluding self-weight of purlin and rafters																	
		Not more than 0.5						More than 0.5 but not more than 0.75						More than 0.75 but not more than 1.00					
		Spacing of purlins (mm)																	
Breadth (mm)	Depth (mm)	1500	1800	2100	2400	2700	3000	1500	1800	2100	2400	2700	3000	1500	1800	2100	2400	2700	3000
		Maximum clear span (m)																	
63	150	2.14	2.00	1.90	1.81			2.00	1.88					1.90					
63	175	2.49	2.34	2.21	2.11	2.02	1.94	2.34	2.19	2.07	1.97	1.88	1.81	2.21	2.07	1.95	1.86		
63	200	2.84	2.67	2.52	2.41	2.30	2.22	2.67	2.50	2.36	2.25	2.15	2.06	2.52	2.36	2.23	2.12	2.01	1.91
63	225	3.20	3.00	2.84	2.70	2.59	2.49	3.00	2.81	2.66	2.53	2.42	2.31	2.84	2.66	2.51	2.39	2.25	2.13
63	275	3.90	3.66	3.46	3.30	3.16	3.04	3.66	3.43	3.24	3.09	2.94	2.79	3.47	3.24	3.07	2.89	2.72	2.57
75	125	1.90																	
75	150	2.27	2.13	2.02	1.92	1.84		2.13	2.00	1.89	1.80			2.02	1.89				
75	175	2.65	2.48	2.35	2.24	2.15	2.07	2.48	2.33	2.20	2.10	2.01	1.93	2.35	2.20	2.08	1.98	1.90	1.83
75	200	3.02	2.84	2.69	2.56	2.46	2.36	2.84	2.66	2.52	2.40	2.30	2.21	2.69	2.52	2.38	2.27	2.17	2.09
75	225	3.40	3.19	3.02	2.88	2.76	2.66	3.19	2.99	2.83	2.70	2.58	2.48	3.02	2.83	2.68	2.55	2.44	2.33

Reproduced from *Span Tables for Solid Timber Members in Floors, Ceilings and Roofs (Excluding Trussed Rafter Roofs) for Dwellings* by permission of TRADA Technology Ltd. (Table 22, p 29)

Table 3.4 Permissible clear spans for common or jack rafters. **Imposed load 0.75 kN/m^2; slope of roof 22.5° or more but less than 30.0°; strength class C16; service class 1 or 2.**

Size of rafter		Dead load (kN/m^2) excluding self-weight of rafter								
		Not more than 0.5			More than 0.5 but not more than 0.75			More than 0.75 but not more than 1.00		
		Spacing of rafters (mm)								
Breadth (mm)	Depth (mm)	400	450	600	400	450	600	400	450	600
		Maximum clear span (m)								
38	100	2.19	2.14	2.02	2.02	1.96	1.83	1.88	1.83	1.69
38	125	2.98	2.87	2.61	2.76	2.68	2.44	2.55	2.47	2.26
38	150	3.57	3.44	3.12	3.35	3.23	2.93	3.18	3.06	2.73
47	100	2.57	2.47	2.24	2.36	2.29	2.10	2.20	2.13	1.96
47	125	3.20	3.08	2.80	3.01	2.89	2.63	2.85	2.74	2.49
47	150	3.83	3.68	3.35	3.60	3.46	3.15	3.41	3.28	2.98
ALS/CLS										
38	89	1.83	1.80	1.70	1.70	1.66	1.55	1.60	1.55	1.44
38	140	3.34	3.21	2.92	3.13	3.01	2.73	2.96	2.86	2.56

Reproduced from *Span Tables for Solid Timber Members in Floors, Ceilings and Roofs (Excluding Trussed Rafter Roofs) for Dwellings* by permission of TRADA Technology Ltd. (Table 20, p 28)

It should be noted that the TRADA tables reprinted here have a pitch range from 22.5° to 30°. However, these tables can be used safely to size timbers for roofs up to 45° pitch, because this results in a slight over sizing of timbers, i.e. erring on the safe side. If ultimate timber economy is of priority, reference should be made to the full TRADA document. Exact comparison is given in the worked example shown later in this text.

HOW DO WE CALCULATE THE LOADING ON THE ROOF?

The dead load, i.e. that of the roof covering itself, is categorised in three groups: a roof covering weighing not more than 0.5 kN/m^2; from 0.5 but not more than 0.75 kN/m^2; and more than 0.75 but not more than 1 kN/m^2. These loadings cover the weight of the tiles and the self weight of the roof structure. The weight of tiles can be found from manufacturers' technical literature, but indications of likely weight can be found in Chapter 8, Figure 8.9. As an indication, interlock tiles generally fall into the category of up to 0.5 kN/m^2, natural slates in the 0.5–0.75 kN/m^2 category, and plain tiles in the category 0.75–1 kN/m^2.*

The imposed load is that for snow lying on the roof; this varies, as has been stated, with geography, altitude and indeed with the roof pitch, but the latter factor is taken into account within the calculations for preparation of the tables. To assess the imposed load, reference should be made to Figure 3.1 (snow roof loading zones) and also to Table 3.5 (imposed snow roof loads for zones defined in Figure 3.1).

TIMBER MEMBER SIZING DESIGN: EXAMPLE

Figure 3.2 illustrates the roof for which timber members are to be designed. The following information is necessary:

The pitch of the roof 45°.
The roof is covered with interlocking tiles.

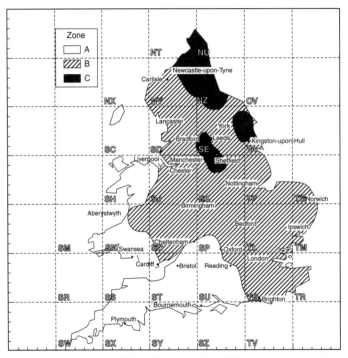

Figure 3.1 Snow roof loading zones. © Building Research Establishment, reproduced with permission.

Calculating size of timber

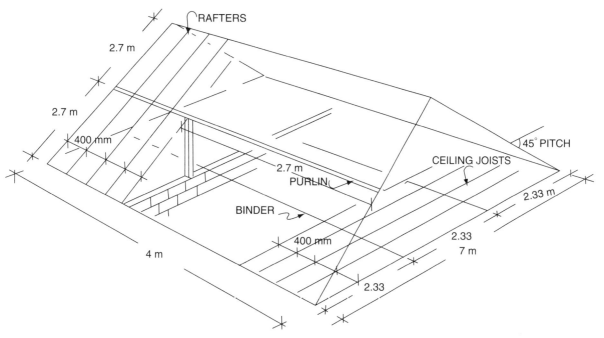

Figure 3.2 Dimensions of roof in worked example.

Table 3.5 Imposed snow roof loads for the zones defined in Figure 3.1. **These data are applicable to small buildings only, as described in BS 6399-3 clause 4.3.2 (1998). For altitudes greater than 200 m refer to guidance given in BS 6399-3.**

Zone	Imposed roof loads (kN/m^2)	
	Altitudes not exceeding 100 m	**Altitudes exceeding 100 m but not exceeding 200 m**
A	0.75	0.75
B	0.75	1.00
C	0.75	Use BS 6399-3

Reproduced from *Span Tables for Solid Timber Members in Floors, Ceilings and Roofs (Excluding Trussed Rafter Roofs) for Dwellings* by permission of TRADA Technology Ltd. (Table 3, p 11)

The roof is located in Nottingham.
The altitude is 60 m (this can be established by locating the site on Google Earth using a PC).
The ceiling joists are to carry normal loft storage only, not flooring loads. The ceiling would be considered as 12 mm plasterboard.

To determine the imposed load on the roof and therefore to decide which table to use, reference should be made to the map in Figure 3.1. It can be seen that Nottingham is in zone B, and from Table 3.5 it can be seen that the imposed snow load for this area is 0.75 kN/m^2. The timber size calculations illustrated in this book refer to that snow loading.

Ceiling Joists
The spacing of the ceiling joists is 400 mm, the span is 2.33 m. With a dead load on the joists of not more than 0.25 kN/m^2, it can be seen from Table 3.1 that a 38 × 120 mm timber section in strength class C16 is safe.

Ceiling Tie Binders
The longest span is 2.7 m between supports for these binders, the spacing is 2.33 m. It can be seen from Table 3.2 that a 63 × 200 mm timber in strength class C16 is safe.

Purlins
Using interlocking tiles, the dead load is in the category not exceeding 0.5 kN/m^2. The snow load is in the category not exceeding 0.75 kN/m^2. The maximum span of the purlin between supports is 2.7 m.

The spacing of the purlins is 2.7 m, and by reference to Table 3.3 it can be seen that this gives either a 63 × 275 mm C16 timber, which has a safe span of 3.16 m, or a 75 × 225 mm C16 timber, which has a safe span of 2.76 m.

Rafters
Dead load is in the category 0.5 kN/m^2. The snow load is not exceeding 0.75 kN/m^2. The rafter spacing is 400 mm centres. The rafter span is 2.7 m maximum. By reference to Table 3.4 this gives a 38 × 125 section C16 timber safe at 2.98 m maximum span, or a 47 × 100 section C16 timber safe at 2.57 m.

The main timber members have now been designed, resulting in a choice of purlin size and a choice of rafter size.

Reference was made earlier to the ultimate economical design of timber size being achieved using the tables quoted in the TRADA document for a 45° pitch: C16 timber or C24 timber (these Tables are not reproduced in this book).

By way of comparison with the sizes calculated above, simply by using the 45° pitch Tables but still with the C16 timber, the purlin would result in a 63 × 275 mm timber member but would be safe at 3.28 m span, or a 75 × 225 mm timber which is safe at 2.86 m span. It can be seen, therefore, that whilst the safe span increases there is no saving on timber section, the next commercially available smaller section not being strong enough in this stress class of timber.

If, however, one chose to use C24 timber, i.e. a higher stress class, then the purlin size would change to 63 × 225 mm or 75 × 200 mm. Some economy is achieved, but at a premium on the cost of the timber itself.

For a further illustration on the rafters, using the correct pitch but using C16 timber, the rafter size would still be 38 × 125 mm; safe at 3.09 m span, or 47 × 100 mm, which is safe at 2.66 m span. In this instance no saving is achieved. However, if one refers to the tables for C24 timber it would be possible to use a 38 × 100 mm timber section or 47 × 100 mm, i.e. the same as the C16 in the latter case.

The thickness of insulation may be a consideration in the case of the design of the rafter, and the depth of the rafter may be governed by the thickness of insulation to be designed in that structure. For instance, with an attic roof it may be desirable to use 150 mm deep rafters and this factor would then influence the width of the rafter designed from the Tables above.

---

*Approximate conversion of kg/m^2 to kN/m^2 equals $\frac{kg/m^2}{100}$ for the tile element of dead load.

4 CALCULATING THE LENGTH AND CUTTING ANGLES OF TIMBER MEMBERS: DATA TABLES 5°–75°

For the purposes of explaining the use of the ready reckoner, reference should be made to the roof constructions illustrated in Figures 4.1 and 4.2. In practical terms these constructions will cover most traditional roof forms and take account of the hip and valley infills used on trussed rafter construction unless fully engineered trussed rafter hip and valleys have been designed. The cutting angles on all timbers for infill rafters on trussed rafter roofs, especially attic designs, can be calculated using the data tables which follow.

Before cutting any of the roof timbers, two vital pieces of information must be known. Firstly the span, i.e. the distance between the outer faces of the wall plate, and secondly the 'run' of the rafter, this being half the span assuming that it is an equally pitched roof with the ridge in the middle of the span. Another vital piece of information is the pitch or the 'rise' of the roof.

Whenever possible the carpenter who is to construct the roof should at least supervise the fixing of the wall plates. These must be straight, level and parallel to each other. Where the roof has to be fitted to a 'T' or 'L' plan form of building, then the carpenter should check that the wall plates of the projections to either side of the main roof are at a true right angle, unless of course designed to be otherwise. Apart from checking overall dimensions with a steel tape, modern laser levels make it quick

Goss's roofing ready reckoner

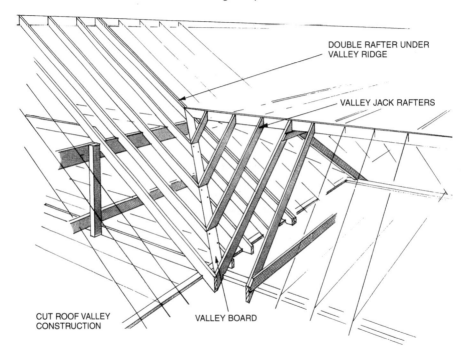

Figure 4.1

Calculating length and cutting angles

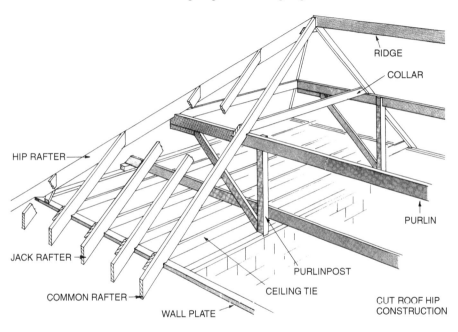

Figure 4.2

and simple to check the level of the plate very effectively, and it is this level of the wall plate which is so important to accurate roof construction.

THE PITCH
The pitch of the roof to be constructed should be clearly stated on the drawings, but if not this should be taken by protractor from the drawings, possibly extending the ceiling line and rafter line away from the point at which they meet, making it easier to get an accurate reading from the protractor. An alternative method to establish the pitch is to calculate the rise of the roof per unit of 'run'. To use the tables in this book, this must be stated in metres rise per metre run. See Figure 4.3.

The Run of the Rafter
The run of the rafter is the horizontal distance covered by the rafter from the wall plate to the ridge. See Figure 4.4.

The Rise of the Rafter
The rise of the rafter is the height from the top of the rafter vertically above the outside of the wall plate, to the top of the rafter at the centre line of the ridge position. See Figure 4.4.

USING THE TABLES TO CUT A COMMON RAFTER
The use of the tables is best explained by a worked example and to do this we will take a roof of pitch at 36° or a rise of 0.727 m per metre run, and a span of 8.46 m. Then the run:

= 8.46 ÷ 2
= 4.23 m

The length of the rafter can now be calculated from the tables referring to 36° pitch. It can be seen that the length of the rafter for 1 m of run = 1.236 m, therefore the length for 4 m of roof:

Calculating length and cutting angles

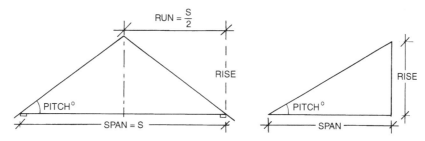

Figure 4.3

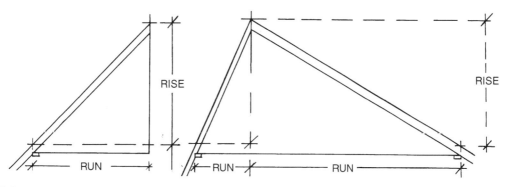

Figure 4.4

= 4 × 1.236
= 4.944 m.

The calculation for the whole rafter length now looks as follows:

4 m	=	4.944 m
0.2 m	=	0.247 m
0.03 m	=	0.0371 m
4.23 m	=	5.2281 m

Now we have the basic length of the rafter.

To calculate the *exact* length of the rafter, the length above must be reduced by $\frac{1}{2}$ the thickness of the ridge; this can be calculated as above if perfection is required. For a ridge thickness of 40 mm, the length of run of the rafter must be reduced by 20 mm; this gives a reduction in rafter length from the tables of 0.0247 m or 24.7 mm. (Tables have to be modified by a factor of 10 because the run above is 20 mm which is 0.02 m and not 0.2 mm as illustrated in the tables.

We then need to add to the rafter the extra length needed to cover the over hang. This can be simply calculated in the same way by finding the length of over hang from the outside of the wall plate to the back of the fascia (see Figure 4.5), and we will assume for the purposes of this calculation that this over hang gives a 450 mm run. Then again by reference to the tables, it will be seen that the additional rafter length required is 556 mm. This now gives an overall rafter length as follows:

Basic rafter		5.2281 m
Add over hang	+	0.556 m
Deduct half ridge	−	0.0247 m
		5.759 m

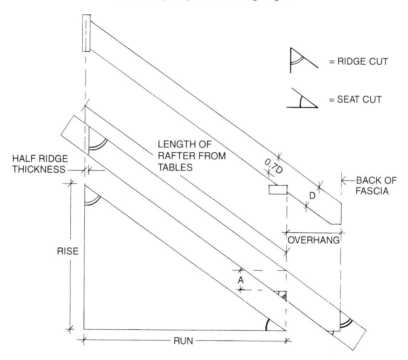

Figure 4.5

Do Not Cut Yet!

Referring now to the tables, mark the ridge bevel on one end for the ridge cut and again at the other end for the fascia cut: in this example the angle will be 54°. The distance between those marks will be the length of the rafter calculated above.

Mark the birdsmouth as shown in Figure 4.5, again using the ridge and seat bevels from the tables as indicated in the illustration.

Next mark the soffit line – this must be taken from the building design. If the soffit cut on the rafter is to be at the soffit line, the seat bevel can again be used. If the soffit line is lower than the lowest point of the rafter, then no soffit cut is required.

Now check all dimensions and angles for this first rafter, which should be regarded as the master.

Now Cut the First Rafter

Using this first rafter check fit to the roof and use this as a pattern to mark out all of the remaining identical rafters. NB to give a true line for the fascia, it is common practice not to cut the fascia cut on all rafters at this stage. Leave the fascia cut on all roof members until the construction is complete, then using a chalked line from one end of the roof to the other, mark the fascia line on the top of the rafters. From this line, using a level, a true plumb line can be marked and the rafter cut. This traditional method involves cutting the fascia cut on the roof itself, which is a time consuming task usually done using a hand saw. Trussed rafter roofs, being prefabricated, generally have the fascia cut made at the factory. If this is the case and allowing for some manufacturing tolerance on span, it will not be possible to line fascia cuts on both sides of the roof. There are two courses of action (1) to re-mark and cut the foot of the trussed rafter again as outlined above, or (2) to use packers to align the fascia onto the pre-cut feet. DO NOT be tempted to align trussed rafters to one side of the roof by aligning their rafter feet. Allowable manufacturing tolerances in a roof of the span we have been discussing could result in a variation of up to 9 mm, thus moving the ridge off the centre line, and up to a 9 mm variance between the feet of the rafters on the opposite side of the roof.

Cutting the common rafter can obviously be done by hand saw, by powered hand saw, or by using a powered compound mitre saw which can be pre-set at the ridge bevel. Then, with a saw table with the length stop at an appropriate position, all cuts will be precisely the same with no further marking required.

HIP JACK RAFTERS

The length of these members will depend upon the centres at which they are to be fixed; by that we mean their spacing centre line to centre line of the thickness of the member, which should match the common rafter spacing. See Figures 2.1 and 4.6.

Continuing with the example above, the basic common rafter length was 5.2281 m, and then assuming a jack rafter spacing of 600 mm by referring to the table, it can be seen that this length must be reduced by 742 mm.

DO NOT forget to add the over hang of the common rafter; DO NOT adjust the jack rafter for the hip until a trial fit has been taken. The jack rafter meets the hip at an angle and must therefore be cut at an angle to meet the hip both horizontally and vertically, giving what is known as a compound cut. The hip, being fixed vertically in its section, gives the same bevel cut at the top of the jack rafter as was used at the ridge and this same ridge bevel can be used. However, the edge cut can be found in the tables as the 'edge bevel', and for this an example can be seen as 39°. With these two angles the top of the hip jack can be marked, and at the lower end, the common rafter master can be used to mark the fascia cut and birdsmouth.

Goss's roofing ready reckoner

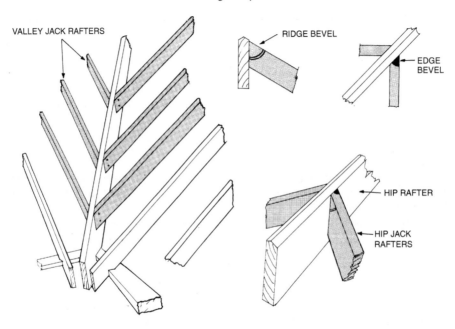

Figure 4.6

HIP RAFTERS

A full hip (i.e. that which is constructed from wall plate to ridge), will have the same rise as the common rafters, and the tables have been calculated on the assumption that the hip is on the mitre of a right angled corner and is therefore at right angles to the common rafters. This allows the same run as the common rafter to be used, to save further calculation.

Continuing with the example above then, the run of the hip would be 4.23 m, and by referring to the tables for the length of hip, the calculation will result in a hip length of 6.7257 m.

The seat and ridge bevels can be taken directly from the tables in this case 27° and 63° respectively. Care must be taken when setting out the birdsmouth to ensure that the depth of rafter (A) illustrated in Figure 4.7, equals that of the common rafter illustrated in Figure 4.5.

The mitre at the top of the hips where they meet the ridge does need the special setting of a bevel. The marking gauge is set to $\frac{1}{2}$ the thickness of the hip and marked on the end of both faces after the plumb cut is made. See Figure 4.7.

Backing of Hips

In a good job, the hips are backed – that is to say a chamfer is planed both ways from the centre line on the top of the hip so that the two surfaces are in line with the planes of the roof on adjacent sides. This gives a good seating for the battens. After cutting the hip to the plumb line the same

Goss's roofing ready reckoner

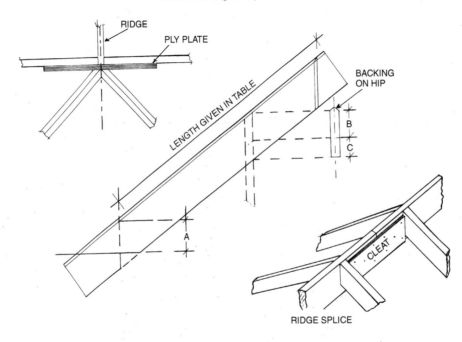

Figure 4.7

plumb bevel for marking the profile of the backing chamfers on each side of the hip may be used. See Figure 4.7.

Dimension B, the length of the plumb cut of the jack rafters is measured on the top end of the hip down from the backing levels, leaving a remainder C. If C is measured along the side bevel of the purlin it gives the position of the projection under the hip. See Figure 4.8.

VALLEY JACK RAFTERS

The tables are based on a construction which assumes a valley rafter similar to the hip rafter; see Figure 4.6, NOT that illustrated in Figure 4.1, which is a more modern construction and one which works with a trussed rafter roof if a prefabricated valley set of frames is not provided. Returning then to the traditional cut valley, this is essentially a hip in reverse. The valley jacks decrease in length as they progress up the roof, and again would be based at similar centres to the common rafters. The same bevels as for the hip jack rafter can be used but this time on the foot of the rafter rather than the ridge as before. The common rafter ridge bevel can be used at the top.

Now referring to Figure 4.1, the top cut on the valley jack is the same as the common rafter, but the foot of the valley rests on a flat valley board nailed on the top of the rafters of the main roof. This construction is suitable for all valleys except attic construction. The cut at the bottom of the valley jack is the same as the seat cut for the common rafter with an edge bevel equal to the pitch of the main roof. This may not be the same as the pitch of the roof on which the valley jack has to be fitted and should therefore be checked.

THE RIDGE

This roof member is usually relatively thin and can be no more than 25–38 mm. It takes little load from the roof as pairs of common rafters or valley rafters are placed directly opposite one another across the ridge; it does act as a tie from gable to gable or hip to hip. In a roof with a gable at both ends, the

Goss's roofing ready reckoner

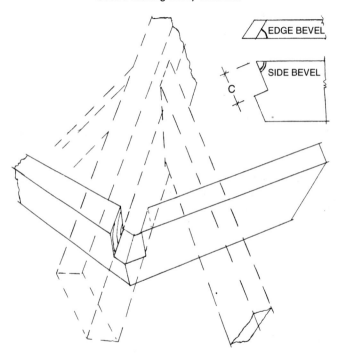

Figure 4.8

length of the ridge equals the length of the wall plate and is usually built into the gables at both ends. In a hip roof with a hip at both ends, the ridge length is normally the length of the building (internally) less the width (internally). If a short packing piece is used at the end of the ridge in a hip construction to give good hip rafter supports, then this thickness must be deducted from the overall length of the ridge for each hip. The packing piece should be at least as deep as the hip rafter ridge cut, and may therefore be too deep for readily available softwood. The use of 19 mm exterior grade plywood is recommended. The ply can also be used to couple ridge pieces together in their length because even the longest lengths of timber may be too short for the overall length of the ridge itself. See Figure 4.7. These 'cleats' should be located between rafters to avoid having to cut back the rafter by the thickness of the cleat.

PURLINS

As illustrated under the definition for purlin earlier in this book, Figure 2.1, the purlin can be fitted either at right angles to the underside of the rafter or vertically. The tables given in this book show edge and side bevels for purlins set at right angles to the rafters. See Figure 4.8. This is the traditional form of construction and may have to be followed on extension work to older buildings.

The purlin fitted vertically is preferred from a structural viewpoint because it acts as a true beam carrying the rafters. The rafters can be better fitted by birdsmouth to the purlin. The edge bevel on this type of purlin is that formed by the hip on plan view which on a right angled hip is 45°. The side bevel is 90°, simply a square cut.

METRIC CALCULATION TABLES

RISE OF COMMON RAFTER 0.087 m **PER METRE OF RUN**   **PITCH** 5°

BEVELS:
- COMMON RAFTER – SEAT 5
- " " – RIDGE 85
- HIP OR VALLEY – SEAT
- " " " – RIDGE
- JACK RAFTER – EDGE
- PURLIN – EDGE
- " – SIDE

JACK RAFTERS 333 mm **CENTRES DECREASE** (in mm to 999 and thereafter in m)
400 " " "
500 " " "
600 " " "

Run of Rafter	0.1	0.2	0.3	0.4	0.5	0.6	0.7	0.8	0.9	1.0
Length of Rafter	0.1	0.201	0.301	0.401	0.502	0.602	0.703	0.803	0.903	1.004
Length of Hip										

Calculating length and cutting angles

RISE OF COMMON RAFTER 0.105 m **PER METRE OF RUN** **PITCH** 6°

BEVELS: COMMON RAFTER – SEAT 6
           ″   ″   – RIDGE 84
         HIP OR VALLEY   – SEAT
           ″   ″   ″   – RIDGE
         JACK RAFTER   – EDGE
         PURLIN   – EDGE
           ″   – SIDE

JACK RAFTERS 333 mm **CENTRES DECREASE**   (in mm to 999 and
                  400 ″   ″   ″   thereafter in m)
                  500 ″   ″   ″
                  600 ″   ″   ″

Run of Rafter	0.1	0.2	0.3	0.4	0.5	0.6	0.7	0.8	0.9	1.0
Length of Rafter	0.101	0.201	0.302	0.402	0.503	0.603	0.704	0.804	0.905	1.006
Length of Hip										

Goss's roofing ready reckoner

RISE OF COMMON RAFTER 0.123 m **PER METRE OF RUN**   **PITCH** 7°

BEVELS: COMMON RAFTER — SEAT 7
                  "   " — RIDGE 83
              HIP OR VALLEY — SEAT
              "   "   " — RIDGE
              JACK RAFTER — EDGE
              PURLIN — EDGE
              " — SIDE

JACK RAFTERS 333 mm **CENTRES DECREASE** (in mm to 999 and
                    400 "   "   "   thereafter in m)
                    500 "   "   "
                    600 "   "   "

Run of Rafter	0.1	0.2	0.3	0.4	0.5	0.6	0.7	0.8	0.9	1.0
Length of Rafter	0.101	0.202	0.302	0.403	0.504	0.605	0.705	0.806	0.907	1.008
Length of Hip										

Calculating length and cutting angles

RISE OF COMMON RAFTER 0.141 m **PER METRE OF RUN**   **PITCH** 8°

BEVELS: COMMON RAFTER — SEAT   8
         "   "   — RIDGE 82
         HIP OR VALLEY — SEAT
         "   "   "   — RIDGE
         JACK RAFTER — EDGE
         PURLIN   — EDGE
         "   — SIDE

JACK RAFTERS 333 mm **CENTRES DECREASE**   (in mm to 999 and
                  400 "   "   "   thereafter in m)
                  500 "   "   "
                  600 "   "   "

Run of Rafter	0.1	0.2	0.3	0.4	0.5	0.6	0.7	0.8	0.9	1.0
Length of Rafter	0.101	0.202	0.303	0.404	0.505	0.606	0.707	0.808	0.909	1.01
Length of Hip										

Goss's roofing ready reckoner

RISE OF COMMON RAFTER 0.158 m **PER METRE OF RUN**　　　**PITCH** 9°

BEVELS:　COMMON RAFTER　– SEAT　9
　　　　　　　　"　　　　"　　　– RIDGE　81
　　　　　HIP OR VALLEY　– SEAT
　　　　　　"　　"　　"　　– RIDGE
　　　　　JACK RAFTER　　– EDGE
　　　　　PURLIN　　　　　　– EDGE
　　　　　　"　　　　　　　　– SIDE

JACK RAFTERS　333 mm **CENTRES DECREASE**　(in mm to 999 and
　　　　　　　　　400　"　　　　"　　　　　"　　　thereafter in m)
　　　　　　　　　500　"　　　　"　　　　　"
　　　　　　　　　600　"　　　　"　　　　　"

Run of Rafter	0.1	0.2	0.3	0.4	0.5	0.6	0.7	0.8	0.9	1.0
Length of Rafter	0.101	0.203	0.304	0.405	0.506	0.608	0.709	0.81	0.911	1.013
Length of Hip										

Calculating length and cutting angles

RISE OF COMMON RAFTER 0.176 m **PER METRE OF RUN**   **PITCH** 10°

BEVELS: COMMON RAFTER   – SEAT   10
         ″   ″   – RIDGE   80
         HIP OR VALLEY   – SEAT
         ″   ″   ″   – RIDGE
         JACK RAFTER   – EDGE
         PURLIN   – EDGE
         ″   – SIDE

JACK RAFTERS 333 mm **CENTRES DECREASE**   (in mm to 999 and
                  400 ″   ″   ″   thereafter in m)
                  500 ″   ″   ″
                  600 ″   ″   ″

Run of Rafter	0.1	0.2	0.3	0.4	0.5	0.6	0.7	0.8	0.9	1.0
Length of Rafter	0.102	0.203	0.305	0.406	0.508	0.609	0.711	0.812	0.914	1.015
Length of Hip										

Goss's roofing ready reckoner

RISE OF COMMON RAFTER 0.194 m **PER METRE OF RUN** **PITCH** 11°

BEVELS: COMMON RAFTER – SEAT 11
                                "   "   – RIDGE 79
                HIP OR VALLEY – SEAT
                  "   "   "   – RIDGE
                JACK RAFTER – EDGE
                PURLIN – EDGE
                    "   – SIDE

JACK RAFTERS 333 mm **CENTRES DECREASE** (in mm to 999 and
                         400 "   "   "   thereafter in m)
                         500 "   "   "
                         600 "   "   "

Run of Rafter	0.1	0.2	0.3	0.4	0.5	0.6	0.7	0.8	0.9	1.0
Length of Rafter	0.102	0.204	0.306	0.407	0.509	0.611	0.713	0.815	0.917	1.019
Length of Hip										

Calculating length and cutting angles

RISE OF COMMON RAFTER 0.213 m **PER METRE OF RUN**   **PITCH** 12°

BEVELS: COMMON RAFTER – SEAT 12
         "   "   – RIDGE 78
         HIP OR VALLEY – SEAT
         "   "   "   – RIDGE
         JACK RAFTER   – EDGE
         PURLIN   – EDGE
         "   – SIDE

JACK RAFTERS 333 mm **CENTRES DECREASE**   (in mm to 999 and
                  400 "   "   "   thereafter in m)
                  500 "   "   "
                  600 "   "   "

Run of Rafter	0.1	0.2	0.3	0.4	0.5	0.6	0.7	0.8	0.9	1.0
Length of Rafter	0.102	0.204	0.307	0.409	0.511	0.613	0.716	0.818	0.92	1.022
Length of Hip										

Goss's roofing ready reckoner

RISE OF COMMON RAFTER 0.231 m PER METRE OF RUN   PITCH 13°

BEVELS: COMMON RAFTER   – SEAT 13
              ″   ″   – RIDGE 77
         HIP OR VALLEY   – SEAT
           ″   ″   ″ – RIDGE
         JACK RAFTER   – EDGE
         PURLIN   – EDGE
              ″   – SIDE

JACK RAFTERS 333 mm **CENTRES DECREASE**   (in mm to 999 and
                  400 ″   ″   ″   thereafter in m)
                  500 ″   ″   ″
                  600 ″   ″   ″

Run of Rafter	0.1	0.2	0.3	0.4	0.5	0.6	0.7	0.8	0.9	1.0
Length of Rafter	0.103	0.205	0.308	0.411	0.513	0.616	0.718	0.821	0.924	1.026
Length of Hip										

Calculating length and cutting angles

RISE OF COMMON RAFTER 0.249 m **PER METRE OF RUN** **PITCH** 14°

BEVELS: COMMON RAFTER — SEAT 14
                           "   "   — RIDGE 76
                HIP OR VALLEY — SEAT
                 "   "   "   — RIDGE
                JACK RAFTER — EDGE
                PURLIN — EDGE
                 "   — SIDE

JACK RAFTERS 333 mm **CENTRES DECREASE** (in mm to 999 and thereafter in m)
                         400 "   "   "
                         500 "   "   "
                         600 "   "   "

Run of Rafter	0.1	0.2	0.3	0.4	0.5	0.6	0.7	0.8	0.9	1.0
Length of Rafter	0.103	0.206	0.309	0.412	0.515	0.618	0.721	0.824	0.928	1.031
Length of Hip										

RISE OF COMMON RAFTER 0.268 m **PER METRE OF RUN** **PITCH** 15°

BEVELS: COMMON RAFTER — SEAT 15
                "   "   — RIDGE 75
              HIP OR VALLEY — SEAT
              "   "   "   — RIDGE
              JACK RAFTER — EDGE
              PURLIN — EDGE
              "   — SIDE

JACK RAFTERS   333 mm **CENTRES DECREASE**   (in mm to 999 and thereafter in m)
                     400 "   "   "
                     500 "   "   "
                     600 "   "   "

Run of Rafter	0.1	0.2	0.3	0.4	0.5	0.6	0.7	0.8	0.9	1.0
Length of Rafter	0.104	0.207	0.311	0.414	0.518	0.621	0.725	0.828	0.932	1.035
Length of Hip										

Calculating length and cutting angles

RISE OF COMMON RAFTER 0.287 m **PER METRE OF RUN**   **PITCH** 16°
(Grecian pitch)

BEVELS:	COMMON RAFTER	– SEAT	16
	" "	– RIDGE	74
	HIP OR VALLEY	– SEAT	11.5
	" " "	– RIDGE	78.5
	JACK RAFTER	– EDGE	44
	PURLIN	– EDGE	46
	"	– SIDE	74.5

JACK RAFTERS 333 mm **CENTRES DECREASE** 346 (in mm to 999 and
　　　　　　　　　400 "　　　　　"　　　　　"　　416　thereafter in m)
　　　　　　　　　500 "　　　　　"　　　　　"　　520
　　　　　　　　　600 "　　　　　"　　　　　"　　624

Run of Rafter	0.1	0.2	0.3	0.4	0.5	0.6	0.7	0.8	0.9	1.0
Length of Rafter	0.104	0.208	0.312	0.416	0.52	0.624	0.728	0.832	0.936	1.04
Length of Hip	0.144	0.289	0.433	0.577	0.722	0.866	1.01	1.154	1.299	1.443

Goss's roofing ready reckoner

RISE OF COMMON RAFTER 0.306 m **PER METRE OF RUN**   **PITCH** 17°

BEVELS:	COMMON RAFTER	– SEAT	17
	" "	– RIDGE	73
	HIP OR VALLEY	– SEAT	12
	" " "	– RIDGE	78
	JACK RAFTER	– EDGE	43.5
	PURLIN	– EDGE	46.5
	"	– SIDE	73.5

JACK RAFTERS	333 mm	**CENTRES DECREASE**	348	(in mm to 999 and
	400 "	" "	418	thereafter in m)
	500 "	" "	522	
	600 "	" "	627	

Run of Rafter	0.1	0.2	0.3	0.4	0.5	0.6	0.7	0.8	0.9	1.0
Length of Rafter	0.105	0.209	0.314	0.418	0.523	0.627	0.732	0.837	0.941	1.046
Length of Hip	0.145	0.289	0.434	0.579	0.723	0.868	1.013	1.158	1.302	1.447

Calculating length and cutting angles

RISE OF COMMON RAFTER 0.315 m **PER METRE OF RUN** **PITCH** $17\frac{1}{2}°$

BEVELS:
	COMMON RAFTER	– SEAT	17.5
	″ ″	– RIDGE	72.5
	HIP OR VALLEY	– SEAT	12.5
	″ ″ ″	– RIDGE	77.5
	JACK RAFTER	– EDGE	43.5
	PURLIN	– EDGE	46.5
	″	– SIDE	73.5

JACK RAFTERS 333 mm **CENTRES DECREASE** 349   (in mm to 999 and
           400 ″   ″   ″   420   thereafter in m)
           500 ″   ″   ″   524
           600 ″   ″   ″   629

Run of Rafter	0.1	0.2	0.3	0.4	0.5	0.6	0.7	0.8	0.9	1.0
Length of Rafter	0.105	0.210	0.315	0.419	0.524	0.629	0.734	0.839	0.944	1.049
Length of Hip	0.145	0.29	0.435	0.579	0.724	0.829	1.014	1.159	1.304	1.449

Goss's roofing ready reckoner

RISE OF COMMON RAFTER 0.325 m PER METRE OF RUN   **PITCH** 18°

	BEVELS:	COMMON RAFTER	– SEAT	18
		″ ″	– RIDGE	72
		HIP OR VALLEY	– SEAT	13
		″ ″ ″	– RIDGE	77
		JACK RAFTER	– EDGE	43.5
		PURLIN	– EDGE	46.5
		″	– SIDE	73

JACK RAFTERS 333 mm **CENTRES DECREASE** 350   (in mm to 999 and
                 400 ″   ″   ″   421   thereafter in m)
                 500 ″   ″   ″   526
                 600 ″   ″   ″   631

Run of Rafter	0.1	0.2	0.3	0.4	0.5	0.6	0.7	0.8	0.9	1.0
Length of Rafter	0.105	0.21	0.315	0.421	0.526	0.631	0.736	0.841	0.946	1.051
Length of Hip	0.145	0.29	0.435	0.58	0.726	0.871	1.016	1.161	1.306	1.451

Calculating length and cutting angles

RISE OF COMMON RAFTER 0.344 m **PER METRE OF RUN**   **PITCH** 19°

BEVELS:	COMMON RAFTER	– SEAT	19
	″ ″	– RIDGE	71
	HIP OR VALLEY	– SEAT	13.5
	″ ″ ″	– RIDGE	76.5
	JACK RAFTER	– EDGE	43.5
	PURLIN	– EDGE	46.5
	″	– SIDE	72

JACK RAFTERS 333 mm **CENTRES DECREASE** 352 (in mm to 999 and
               400 ″   ″   ″   423   thereafter in m)
               500 ″   ″   ″   529
               600 ″   ″   ″   635

Run of Rafter	0.1	0.2	0.3	0.4	0.5	0.6	0.7	0.8	0.9	1.0
Length of Rafter	0.106	0.212	0.317	0.423	0.529	0.635	0.74	0.846	0.952	1.058
Length of Hip	0.146	0.291	0.437	0.582	0.728	0.873	1.019	1.164	1.31	1.456

RISE OF COMMON RAFTER 0.364 m **PER METRE OF RUN**   **PITCH** 20°

BEVELS: COMMON RAFTER   – SEAT   20
                     ″   ″   – RIDGE   70
             HIP OR VALLEY   – SEAT   14.5
             ″   ″   ″   – RIDGE   75.5
             JACK RAFTER   – EDGE   43
             PURLIN   – EDGE   47
                 ″   – SIDE   71

JACK RAFTERS   333 mm   **CENTRES DECREASE**   354   (in mm to 999 and
                       400   ″   ″   ″   426   thereafter in m)
                       500   ″   ″   ″   532
                       600   ″   ″   ″   638

Run of Rafter	0.1	0.2	0.3	0.4	0.5	0.6	0.7	0.8	0.9	1.0
Length of Rafter	0.106	0.213	0.319	0.426	0.532	0.639	0.745	0.851	0.958	1.064
Length of Hip	0.146	0.292	0.438	0.584	0.73	0.876	1.022	1.168	1.314	1.46

Calculating length and cutting angles

RISE OF COMMON RAFTER 0.384 m **PER METRE OF RUN**   **PITCH** 21°

BEVELS: COMMON RAFTER   – SEAT   21
                       ″   ″   – RIDGE   69
               HIP OR VALLEY   – SEAT   15
                ″   ″   ″   – RIDGE   75
               JACK RAFTER   – EDGE   43
               PURLIN   – EDGE   47
                  ″   – SIDE   70.5

JACK RAFTERS   333 mm   **CENTRES DECREASE**   357   (in mm to 999 and
                          400   ″   ″   ″   428   thereafter in m)
                          500   ″   ″   ″   536
                          600   ″   ″   ″   643

Run of Rafter	0.1	0.2	0.3	0.4	0.5	0.6	0.7	0.8	0.9	1.0
Length of Rafter	0.107	0.214	0.321	0.428	0.536	0.643	0.75	0.857	0.964	1.071
Length of Hip	0.147	0.293	0.44	0.586	0.733	0.879	1.026	1.172	1.319	1.465

Goss's roofing ready reckoner

RISE OF COMMON RAFTER 0.404 m **PER METRE OF RUN**   **PITCH** 22°

BEVELS: COMMON RAFTER   – SEAT   22
                  ″   ″   – RIDGE 68
              HIP OR VALLEY   – SEAT   16
              ″   ″   ″   – RIDGE 74
              JACK RAFTER   – EDGE   43
              PURLIN   – EDGE   47
                  ″   – SIDE   69.5

JACK RAFTERS 333 mm **CENTRES DECREASE** 359   (in mm to 999 and
                                400 ″   ″   ″   432   thereafter in m)
                                500 ″   ″   ″   540
                                600 ″   ″   ″   647

Run of Rafter	0.1	0.2	0.3	0.4	0.5	0.6	0.7	0.8	0.9	1.0
Length of Rafter	0.108	0.216	0.324	0.431	0.539	0.647	0.755	0.863	0.971	1.079
Length of Hip	0.147	0.294	0.441	0.588	0.736	0.883	1.03	1.177	1.324	1.471

Calculating length and cutting angles

RISE OF COMMON RAFTER 0.414 m **PER METRE OF RUN** **PITCH** 22½°

BEVELS:	COMMON RAFTER	– SEAT	22.5
	″ ″	– RIDGE	67.5
	HIP OR VALLEY	– SEAT	16.25
	″ ″ ″	– RIDGE	73.75
	JACK RAFTER	– EDGE	42.75
	PURLIN	– EDGE	47.5
	″	– SIDE	69.0

JACK RAFTERS 333 mm **CENTRES DECREASE** 361 (in mm to 999 and
400 ″   ″   ″   433   thereafter in m)
500 ″   ″   ″   542
600 ″   ″   ″   650

Run of Rafter	0.1	0.2	0.3	0.4	0.5	0.6	0.7	0.8	0.9	1.0
Length of Rafter	0.108	0.216	0.325	0.433	0.541	0.649	0.758	0.866	0.974	1.082
Length of Hip	0.147	0.294	0.442	0.589	0.737	0.884	1.032	1.179	1.326	1.473

Goss's roofing ready reckoner

RISE OF COMMON RAFTER 0.424 m PER METRE OF RUN   **PITCH** 23°

BEVELS: COMMON RAFTER   – SEAT   23
         ″   ″   – RIDGE 67
         HIP OR VALLEY   – SEAT   16.5
         ″   ″   ″   – RIDGE 73.5
         JACK RAFTER   – EDGE   42.5
         PURLIN   – EDGE   47.5
         ″   – SIDE   68.5

JACK RAFTERS 333 mm **CENTRES DECREASE** 362   (in mm to 999 and
                  400 ″   ″   ″   434   thereafter in m)
                  500 ″   ″   ″   543
                  600 ″   ″   ″   652

Run of Rafter	0.1	0.2	0.3	0.4	0.5	0.6	0.7	0.8	0.9	1.0
Length of Rafter	0.109	0.217	0.326	0.435	0.543	0.652	0.76	0.869	0.978	1.086
Length of Hip	0.148	0.295	0.443	0.591	0.739	0.886	1.034	1.181	1.329	1.477

Calculating length and cutting angles

RISE OF COMMON RAFTER   0.445 m PER METRE OF RUN   **PITCH** 24°
(Roman pitch)

BEVELS:	COMMON RAFTER	– SEAT	24
	" "	– RIDGE	66
	HIP OR VALLEY	– SEAT	17.5
	" " "	– RIDGE	72.5
	JACK RAFTER	– EDGE	42.5
	PURLIN	– EDGE	47.5
	"	– SIDE	68

JACK RAFTERS 333 mm **CENTRES DECREASE** 365   (in mm to 999 and
             400 "   "   "   438   thereafter in m)
             500 "   "   "   548
             600 "   "   "   657

Run of Rafter	0.1	0.2	0.3	0.4	0.5	0.6	0.7	0.8	0.9	1.0
Length of Rafter	0.109	0.219	0.328	0.438	0.547	0.657	0.766	0.876	0.985	1.095
Length of Hip	0.148	0.297	0.445	0.593	0.741	0.89	1.038	1.186	1.334	1.483

RISE OF COMMON RAFTER 0.466 m **PER METRE OF RUN**   **PITCH** 25°

BEVELS:	COMMON RAFTER	– SEAT	25
	" "	– RIDGE	65
	HIP OR VALLEY	– SEAT	18
	" " "	– RIDGE	72
	JACK RAFTER	– EDGE	42
	PURLIN	– EDGE	48
	"	– SIDE	67

JACK RAFTERS 333 mm **CENTRES DECREASE** 367   (in mm to 999 and
             400 "   "   "   441   thereafter in m)
             500 "   "   "   552
             600 "   "   "   662

Run of Rafter	0.1	0.2	0.3	0.4	0.5	0.6	0.7	0.8	0.9	1.0
Length of Rafter	0.11	0.221	0.331	0.441	0.552	0.662	0.772	0.883	0.993	1.103
Length of Hip	0.149	0.298	0.447	0.596	0.745	0.893	1.042	1.191	1.34	1.489

Calculating length and cutting angles

RISE OF COMMON RAFTER 0.5 m **PER METRE OF RUN**   **PITCH** 26° 34′
(Quarter pitch)

BEVELS: COMMON RAFTER   – SEAT   26.5
         ″   ″   – RIDGE 63.5
         HIP OR VALLEY   – SEAT   19.5
         ″   ″   ″   – RIDGE 70.5
         JACK RAFTER   – EDGE   42
         PURLIN   – EDGE   48
         ″   – SIDE   66

JACK RAFTERS 333 mm **CENTRES DECREASE** 372 (in mm to 999 and
                 400 ″   ″   ″ 447   thereafter in m)
                 500 ″   ″   ″ 559
                 600 ″   ″   ″ 671

Run of Rafter	0.1	0.2	0.3	0.4	0.5	0.6	0.7	0.8	0.9	1.0
Length of Rafter	0.112	0.224	0.335	0.447	0.559	0.671	0.783	0.894	1.006	1.118
Length of Hip	0.15	0.3	0.45	0.6	0.75	0.9	1.05	1.2	1.35	1.5

Goss's roofing ready reckoner

RISE OF COMMON RAFTER 0.521 m **PER METRE OF RUN** **PITCH** $27\frac{1}{2}°$

	BEVELS:	COMMON RAFTER	– SEAT	27.5
		" "	– RIDGE	62.5
		HIP OR VALLEY	– SEAT	20
		" " "	– RIDGE	70
		JACK RAFTER	– EDGE	41.75
		PURLIN	– EDGE	48.5
		"	– SIDE	65

JACK RAFTERS 333 mm **CENTRES DECREASE** 375   (in mm to 999 and
400 " " " 451   thereafter in m)
500 " " " 563
600 " " " 677

Run of Rafter	0.1	0.2	0.3	0.4	0.5	0.6	0.7	0.8	0.9	1.0
Length of Rafter	0.113	0.225	0.338	0.451	0.564	0.676	0.789	0.902	1.015	1.127
Length of Hip	0.151	0.301	0.452	0.603	0.754	0.904	1.054	1.206	1.365	1.507

Calculating length and cutting angles

RISE OF COMMON RAFTER 0.532 m **PER METRE OF RUN**   **PITCH** 28°

	BEVELS:	COMMON RAFTER	– SEAT	28
		" "	– RIDGE	62
		HIP OR VALLEY	– SEAT	20.5
		" " "	– RIDGE	69.5
		JACK RAFTER	– EDGE	41.5
		PURLIN	– EDGE	48.5
		"	– SIDE	65

JACK RAFTERS 333 mm **CENTRES DECREASE** 377 (in mm to 999 and
                 400 "   "   "   453   thereafter in m)
                 500 "   "   "   566
                 600 "   "   "   680

Run of Rafter	0.1	0.2	0.3	0.4	0.5	0.6	0.7	0.8	0.9	1.0
Length of Rafter	0.113	0.227	0.34	0.453	0.566	0.68	0.793	0.906	1.019	1.133
Length of Hip	0.151	0.302	0.453	0.603	0.754	0.905	1.056	1.207	1.358	1.511

Goss's roofing ready reckoner

RISE OF COMMON RAFTER 0.544 m **PER METRE OF RUN**   **PITCH** 29°

BEVELS: COMMON RAFTER – SEAT 29
              "   "   – RIDGE 61
         HIP OR VALLEY   – SEAT 21.5
           "   "   "   – RIDGE 68.5
         JACK RAFTER   – EDGE 41
         PURLIN   – EDGE 49
             "   – SIDE 64

JACK RAFTERS 333 mm **CENTRES DECREASE** 381   (in mm to 999 and
                  400 "   "   "   457   thereafter in m)
                  500 "   "   "   572
                  600 "   "   "   686

Run of Rafter	0.1	0.2	0.3	0.4	0.5	0.6	0.7	0.8	0.9	1.0
Length of Rafter	0.114	0.229	0.343	0.457	0.572	0.686	0.8	0.914	1.029	1.143
Length of Hip	0.152	0.304	0.456	0.608	0.759	0.912	1.063	1.215	1.367	1.519

Calculating length and cutting angles

RISE OF COMMON RAFTER 0.577 m **PER METRE OF RUN**   **PITCH** 30°

BEVELS:	COMMON RAFTER	– SEAT	30
	″ ″	– RIDGE	60
	HIP OR VALLEY	– SEAT	22
	″ ″ ″	– RIDGE	68
	JACK RAFTER	– EDGE	41
	PURLIN	– EDGE	49
	″	– SIDE	63.5

JACK RAFTERS	333 mm	**CENTRES DECREASE**	385	(in mm to 999 and
	400 ″	″ ″	462	thereafter in m)
	500 ″	″ ″	577	
	600 ″	″ ″	693	

Run of Rafter	0.1	0.2	0.3	0.4	0.5	0.6	0.7	0.8	0.9	1.0
Length of Rafter	0.116	0.231	0.346	0.462	0.577	0.693	0.808	0.924	1.039	1.155
Length of Hip	0.153	0.306	0.458	0.611	0.764	0.917	1.069	1.222	1.375	1.528

RISE OF COMMON RAFTER 0.601 m **PER METRE OF RUN** **PITCH** 31°

BEVELS: COMMON RAFTER – SEAT 31
   "   "   – RIDGE 59
  HIP OR VALLEY – SEAT 23
   "   "   "   – RIDGE 67
  JACK RAFTER   – EDGE 40.5
  PURLIN   – EDGE 49.5
   "   – SIDE 62.5

JACK RAFTERS 333 mm **CENTRES DECREASE** 389 (in mm to 999 and
 400 "   "   "   467 thereafter in m)
 500 "   "   "   584
 600 "   "   "   700

Run of Rafter	0.1	0.2	0.3	0.4	0.5	0.6	0.7	0.8	0.9	1.0
Length of Rafter	0.117	0.233	0.35	0.467	0.583	0.7	0.817	0.933	1.05	1.167
Length of Hip	0.154	0.307	0.461	0.615	0.768	0.922	1.076	1.229	1.383	1.537

Calculating length and cutting angles

RISE OF COMMON RAFTER 0.625 m **PER METRE OF RUN**   **PITCH** 32°

BEVELS: COMMON RAFTER – SEAT 32
  "   "   – RIDGE 58
  HIP OR VALLEY   – SEAT 24
  "   "   "   – RIDGE 66
  JACK RAFTER   – EDGE 40.5
  PURLIN   – EDGE 49.5
  "   – SIDE 62

JACK RAFTERS 333 mm **CENTRES DECREASE** 393 (in mm to 999 and
                 400 "   "   " 472 thereafter in m)
                 500 "   "   " 590
                 600 "   "   " 707

Run of Rafter	0.1	0.2	0.3	0.4	0.5	0.6	0.7	0.8	0.9	1.0
Length of Rafter	0.118	0.239	0.354	0.472	0.59	0.708	0.825	0.943	1.061	1.179
Length of Hip	0.155	0.309	0.464	0.618	0.773	0.928	1.082	1.237	1.391	1.546

RISE OF COMMON RAFTER 0.637 m **PER METRE OF RUN**   **PITCH** $32\frac{1}{2}°$

	BEVELS:	COMMON RAFTER	– SEAT	32.5
		" "	– RIDGE	57.5
		HIP OR VALLEY	– SEAT	24.25
		" " "	– RIDGE	65.75
		JACK RAFTER	– EDGE	40.25
		PURLIN	– EDGE	49.75
		"	– SIDE	61.5

JACK RAFTERS   333 mm   **CENTRES DECREASE**   395   (in mm to 999 and
                      400 "   "   "   475   thereafter in m)
                      500 "   "   "   593
                      600 "   "   "   711

Run of Rafter	0.1	0.2	0.3	0.4	0.5	0.6	0.7	0.8	0.9	1.0
Length of Rafter	0.119	0.237	0.356	0.474	0.593	0.711	0.830	0.949	1.067	1.186
Length of Hip	0.155	0.310	0.466	0.620	0.776	0.930	1.086	1.241	1.396	1.551

Calculating length and cutting angles

RISE OF COMMON RAFTER 0.649 m **PER METRE OF RUN**   **PITCH** 33°

BEVELS: COMMON RAFTER – SEAT   33
          ″   ″   – RIDGE 57
         HIP OR VALLEY   – SEAT   24.5
          ″   ″   ″   – RIDGE 65.5
         JACK RAFTER   – EDGE   40
         PURLIN   – EDGE   50
          ″   – SIDE   61.5

JACK RAFTERS 333 mm **CENTRES DECREASE** 397 (in mm to 999 and
                 400 ″   ″   ″   477   thereafter in m)
                 500 ″   ″   ″   596
                 600 ″   ″   ″   715

Run of Rafter	0.1	0.2	0.3	0.4	0.5	0.6	0.7	0.8	0.9	1.0
Length of Rafter	0.119	0.238	0.358	0.48	0.596	0.715	0.835	0.954	1.073	1.192
Length of Hip	0.156	0.311	0.467	0.623	0.778	0.934	1.089	1.245	1.401	1.556

Goss's roofing ready reckoner

RISE OF COMMON RAFTER 0.666 m **PER METRE OF RUN** **PITCH** 33° 40′
(Third pitch)

BEVELS: COMMON RAFTER – SEAT 33.5
         ″   ″   – RIDGE 56.5
         HIP OR VALLEY   – SEAT 25
         ″   ″   ″   – RIDGE 65
         JACK RAFTER   – EDGE 40
         PURLIN   – EDGE 50
         ″   – SIDE 61

JACK RAFTERS 333 mm **CENTRES DECREASE** 397 (in mm to 999 and
                 400 ″   ″   ″   481 thereafter in m)
                 500 ″   ″   ″   601
                 600 ″   ″   ″   721

Run of Rafter	0.1	0.2	0.3	0.4	0.5	0.6	0.7	0.8	0.9	1.0
Length of Rafter	0.12	0.24	0.361	0.481	0.601	0.721	0.841	0.961	1.082	1.202
Length of Hip	0.157	0.313	0.47	0.626	0.782	0.938	1.094	1.251	1.408	1.563

Calculating length and cutting angles

RISE OF COMMON RAFTER 0.7 m **PER METRE OF RUN** **PITCH** 35°

BEVELS: COMMON RAFTER   – SEAT   35
                  ″   ″   – RIDGE 55
          HIP OR VALLEY   – SEAT   26.5
              ″   ″   ″   – RIDGE 63.5
          JACK RAFTER   – EDGE   39.5
          PURLIN   – EDGE   50.5
              ″   – SIDE   60

JACK RAFTERS 333 mm **CENTRES DECREASE** 407   (in mm to 999 and
                  400 ″   ″   ″   488   thereafter in m)
                  500 ″   ″   ″   611
                  600 ″   ″   ″   733

Run of Rafter	0.1	0.2	0.3	0.4	0.5	0.6	0.7	0.8	0.9	1.0
Length of Rafter	0.122	0.244	0.366	0.488	0.61	0.732	0.855	0.977	1.099	1.221
Length of Hip	0.158	0.316	0.473	0.631	0.789	0.947	1.105	1.262	1.42	1.578

RISE OF COMMON RAFTER 0.727 m **PER METRE OF RUN**   **PITCH** 36°

BEVELS:	COMMON RAFTER	– SEAT	36
	" "	– RIDGE	54
	HIP OR VALLEY	– SEAT	27
	" " "	– RIDGE	63
	JACK RAFTER	– EDGE	39
	PURLIN	– EDGE	51
	"	– SIDE	59.5

JACK RAFTERS 333 mm **CENTRES DECREASE** 412 (in mm to 999 and
            400 "   "   "   494   thereafter in m)
            500 "   "   "   618
            600 "   "   "   742

Run of Rafter	0.1	0.2	0.3	0.4	0.5	0.6	0.7	0.8	0.9	1.0
Length of Rafter	0.124	0.247	0.371	0.494	0.618	0.742	0.865	0.989	1.112	1.236
Length of Hip	0.159	0.318	0.477	0.636	0.795	0.954	1.113	1.272	1.431	1.59

Calculating length and cutting angles

RISE OF COMMON RAFTER 0.754 m **PER METRE OF RUN**   **PITCH** 37°

BEVELS: COMMON RAFTER — SEAT 37
         "   "   — RIDGE 53
         HIP OR VALLEY — SEAT 28
         "   "   "   — RIDGE 62
         JACK RAFTER   — EDGE 38.5
         PURLIN   — EDGE 51.5
         "   — SIDE 59

JACK RAFTERS 333 mm **CENTRES DECREASE** 417 (in mm to 999 and
                 400 "   "   "   501   thereafter in m)
                 500 "   "   "   626
                 600 "   "   "   751

Run of Rafter	0.1	0.2	0.3	0.4	0.5	0.6	0.7	0.8	0.9	1.0
Length of Rafter	0.125	0.25	0.376	0.501	0.626	0.751	0.876	1.002	1.127	1.252
Length of Hip	0.16	0.32	0.481	0.641	0.801	0.961	1.122	1.282	1.442	1.602

Goss's roofing ready reckoner

RISE OF COMMON RAFTER 0.767 m **PER METRE OF RUN**   **PITCH** 37½°

BEVELS:	COMMON RAFTER	– SEAT	37.5
	" "	– RIDGE	52.5
	HIP OR VALLEY	– SEAT	28.5
	" " "	– RIDGE	61.5
	JACK RAFTER	– EDGE	38.25
	PURLIN	– EDGE	51.75
	"	– SIDE	58.5

JACK RAFTERS	333 mm	**CENTRES DECREASE**	420	(in mm to 999 and
	400 "	" "	505	thereafter in m)
	500 "	" "	630	
	600 "	" "	757	

Run of Rafter	0.1	0.2	0.3	0.4	0.5	0.6	0.7	0.8	0.9	1.0
Length of Rafter	0.126	0.252	0.378	0.504	0.630	0.756	0.882	1.008	1.134	1.260
Length of Hip	0.161	0.322	0.482	0.636	0.804	1.965	1.126	1.286	1.448	1.609

Calculating length and cutting angles

RISE OF COMMON RAFTER 0.781 m **PER METRE OF RUN** **PITCH** 38°

BEVELS: COMMON RAFTER – SEAT 38
                             ″   ″   – RIDGE 52
            HIP OR VALLEY – SEAT 29
             ″   ″   ″   – RIDGE 61
            JACK RAFTER – EDGE 38
            PURLIN – EDGE 52
             ″   – SIDE 58.5

JACK RAFTERS 333 mm **CENTRES DECREASE** 423 (in mm to 999 and
                      400 ″   ″   ″   508 thereafter in m)
                      500 ″   ″   ″   635
                      600 ″   ″   ″   761

Run of Rafter	0.1	0.2	0.3	0.4	0.5	0.6	0.7	0.8	0.9	1.0
Length of Rafter	0.127	0.254	0.381	0.508	0.635	0.761	0.888	1.015	1.142	1.269
Length of Hip	0.162	0.323	0.485	0.646	0.808	0.969	1.131	1.293	1.454	1.616

RISE OF COMMON RAFTER 0.81 m **PER METRE OF RUN**   **PITCH** 39°

BEVELS:
- COMMON RAFTER — SEAT 39
-     ″   ″ — RIDGE 51
- HIP OR VALLEY — SEAT 30
- ″ ″ ″ — RIDGE 60
- JACK RAFTER — EDGE 38
- PURLIN — EDGE 52
-   ″ — SIDE 58

JACK RAFTERS 333 mm **CENTRES DECREASE** 429   (in mm to 999 and
                                  400 ″   ″   ″   515   thereafter in m)
                                  500 ″   ″   ″   644
                                  600 ″   ″   ″   772

Run of Rafter	0.1	0.2	0.3	0.4	0.5	0.6	0.7	0.8	0.9	1.0
Length of Rafter	0.129	0.257	0.386	0.515	0.643	0.772	0.901	1.029	1.158	1.287
Length of Hip	0.163	0.326	0.489	0.652	0.815	0.978	1.141	1.304	1.467	1.63

Calculating length and cutting angles

RISE OF COMMON RAFTER 0.839 m **PER METRE OF RUN**   **PITCH** 40°

BEVELS:	COMMON RAFTER	– SEAT	40
	" "	– RIDGE	50
	HIP OR VALLEY	– SEAT	30.5
	" " "	– RIDGE	59.5
	JACK RAFTER	– EDGE	37.5
	PURLIN	– EDGE	52.5
	"	– SIDE	57.5

JACK RAFTERS 333 mm **CENTRES DECREASE** 435   (in mm to 999 and
  400 "   "   "   522   thereafter in m)
  500 "   "   "   653
  600 "   "   "   783

Run of Rafter	0.1	0.2	0.3	0.4	0.5	0.6	0.7	0.8	0.9	1.0
Length of Rafter	0.131	0.261	0.392	0.522	0.653	0.783	0.914	1.044	1.175	1.305
Length of Hip	0.164	0.329	0.493	0.658	0.822	0.987	1.151	1.316	1.48	1.644

Goss's roofing ready reckoner

RISE OF COMMON RAFTER 0.869 m **PER METRE OF RUN** **PITCH** 41°

BEVELS:	COMMON RAFTER	– SEAT	41
	" "	– RIDGE	49
	HIP OR VALLEY	– SEAT	31.5
	" " "	– RIDGE	58.5
	JACK RAFTER	– EDGE	37
	PURLIN	– EDGE	53
	"	– SIDE	56.5

JACK RAFTERS 333 mm **CENTRES DECREASE** 441 (in mm to 999 and
  400 "   "   "   530   thereafter in m)
  500 "   "   "   663
  600 "   "   "   795

Run of Rafter	0.1	0.2	0.3	0.4	0.5	0.6	0.7	0.8	0.9	1.0
Length of Rafter	0.133	0.265	0.398	0.53	0.663	0.795	0.928	1.06	1.193	1.325
Length of Hip	0.166	0.332	0.498	0.664	0.83	0.996	1.162	1.328	1.494	1.66

Calculating length and cutting angles

RISE OF COMMON RAFTER 0.9 m PER METRE OF RUN   PITCH 42°

BEVELS: COMMON RAFTER — SEAT 42
           ″   ″   — RIDGE 48
         HIP OR VALLEY   — SEAT 32.5
           ″   ″   ″   — RIDGE 57.5
         JACK RAFTER   — EDGE 36.5
         PURLIN   — EDGE 53.5
           ″   — SIDE 56

JACK RAFTERS 333 mm **CENTRES DECREASE** 448   (in mm to 999 and
                  400 ″   ″   ″   538   thereafter in m)
                  500 ″   ″   ″   673
                  600 ″   ″   ″   808

Run of Rafter	0.1	0.2	0.3	0.4	0.5	0.6	0.7	0.8	0.9	1.0
Length of Rafter	0.135	0.269	0.404	0.538	0.673	0.807	0.942	1.097	1.211	1.346
Length of Hip	0.168	0.335	0.503	0.671	0.838	1.006	1.173	1.341	1.509	1.677

Goss's roofing ready reckoner

RISE OF COMMON RAFTER 0.916 m **PER METRE OF RUN**   **PITCH** 42½°

BEVELS: COMMON RAFTER – SEAT   42.5
         "   "   – RIDGE 47.5
         HIP OR VALLEY – SEAT   33
         "   "   "   – RIDGE 57
         JACK RAFTER   – EDGE   36.25
         PURLIN   – EDGE   53.75
              "   – SIDE   55.75

JACK RAFTERS 333 mm **CENTRES DECREASE** 452   (in mm to 999 and
                  400 "   "   "   543   thereafter in m)
                  500 "   "   "   679
                  600 "   "   "   815

Run of Rafter	0.1	0.2	0.3	0.4	0.5	0.6	0.7	0.8	0.9	1.0
Length of Rafter	0.137	0.271	0.406	0.543	0.678	0.814	0.949	1.085	1.221	1.356
Length of Hip	0.170	0.337	0.505	0.674	0.842	1.011	1.179	1.348	1.569	1.685

Calculating length and cutting angles

RISE OF COMMON RAFTER 0.933 m **PER METRE OF RUN**   **PITCH** 43°

BEVELS:	COMMON RAFTER	– SEAT	43
	" "	– RIDGE	47
	HIP OR VALLEY	– SEAT	33.5
	" " "	– RIDGE	56.5
	JACK RAFTER	– EDGE	36
	PURLIN	– EDGE	54
	"	– SIDE	55.5

JACK RAFTERS 333 mm **CENTRES DECREASE** 455   (in mm to 999 and
             400 "   "   "   547   thereafter in m)
             500 "   "   "   684
             600 "   "   "   820

Run of Rafter	0.1	0.2	0.3	0.4	0.5	0.6	0.7	0.8	0.9	1.0
Length of Rafter	0.137	0.273	0.41	0.547	0.684	0.82	0.967	1.094	1.231	1.367
Length of Hip	0.169	0.339	0.508	0.678	0.847	1.016	1.186	1.355	1.525	1.694

Goss's roofing ready reckoner

RISE OF COMMON RAFTER 0.966 m **PER METRE OF RUN**   **PITCH** 44°

	BEVELS:	COMMON RAFTER	– SEAT	44
		″ ″	– RIDGE	46
		HIP OR VALLEY	– SEAT	34.5
		″ ″ ″	– RIDGE	55.5
		JACK RAFTER	– EDGE	35.5
		PURLIN	– EDGE	54.5
		″	– SIDE	55

JACK RAFTERS 333 mm **CENTRES DECREASE** 463   (in mm to 999 and
                 400 ″   ″   ″   556   thereafter in m)
                 500 ″   ″   ″   695
                 600 ″   ″   ″   834

Run of Rafter	0.1	0.2	0.3	0.4	0.5	0.6	0.7	0.8	0.9	1.0
Length of Rafter	0.139	0.278	0.417	0.556	0.695	0.834	0.973	1.111	1.251	1.39
Length of Hip	0.171	0.342	0.514	0.685	0.856	1.027	1.199	1.37	1.541	1.712

Calculating length and cutting angles

RISE OF COMMON RAFTER 1.0 m PER METRE OF RUN   PITCH 45°

BEVELS: COMMON RAFTER − SEAT   45
         "   "   − RIDGE 45
         HIP OR VALLEY   − SEAT   35.5
         "   "   "   − RIDGE 54.5
         JACK RAFTER   − EDGE   35.5
         PURLIN   − EDGE   54.5
         "   − SIDE   54.5

JACK RAFTERS 333 mm **CENTRES DECREASE** 471   (in mm to 999 and
                 400 "   "   "   566   thereafter in m)
                 500 "   "   "   707
                 600 "   "   "   848

Run of Rafter	0.1	0.2	0.3	0.4	0.5	0.6	0.7	0.8	0.9	1.0
Length of Rafter	0.141	0.283	0.424	0.566	0.707	0.848	0.99	1.131	1.273	1.414
Length of Hip	0.173	0.346	0.519	0.693	0.866	1.039	1.212	1.386	1.559	1.732

Goss's roofing ready reckoner

RISE OF COMMON RAFTER 1.036 m **PER METRE OF RUN**   **PITCH** 46°

BEVELS:
- COMMON RAFTER — SEAT 46
-        "   "   — RIDGE 44
- HIP OR VALLEY — SEAT 36
-    "   "   "   — RIDGE 54
- JACK RAFTER — EDGE 35
- PURLIN — EDGE 55
-    "   — SIDE 54.5

JACK RAFTERS 333 mm **CENTRES DECREASE** 480   (in mm to 999 and
                            400 "   "   "   576   thereafter in m)
                            500 "   "   "   720
                            600 "   "   "   864

Run of Rafter	0.1	0.2	0.3	0.4	0.5	0.6	0.7	0.8	0.9	1.0
Length of Rafter	0.144	0.288	0.432	0.576	0.72	0.864	1.058	1.152	1.296	1.44
Length of Hip	0.175	0.351	0.526	0.701	0.876	1.052	1.227	1.402	1.578	1.753

Calculating length and cutting angles

RISE OF COMMON RAFTER 1.072 m **PER METRE OF RUN**   **PITCH** 47°

BEVELS: COMMON RAFTER   – SEAT   47
              ″   ″   – RIDGE   43
         HIP OR VALLEY   – SEAT   37
              ″   ″   ″   – RIDGE   53
         JACK RAFTER   – EDGE   34.5
         PURLIN   – EDGE   55.5
              ″   – SIDE   54

JACK RAFTERS 333 mm **CENTRES DECREASE** 488   (in mm to 999 and
                  400 ″   ″   ″   586   thereafter in m)
                  500 ″   ″   ″   733
                  600 ″   ″   ″   880

Run of Rafter	0.1	0.2	0.3	0.4	0.5	0.6	0.7	0.8	0.9	1.0
Length of Rafter	0.147	0.293	0.44	0.587	0.733	0.88	1.026	1.173	1.32	1.466
Length of Hip	0.177	0.355	0.532	0.71	0.887	1.065	1.242	1.42	1.597	1.775

RISE OF COMMON RAFTER 1.111 m **PER METRE OF RUN**   **PITCH** 48°

BEVELS: COMMON RAFTER – SEAT 48
         "   "   – RIDGE 42
         HIP OR VALLEY – SEAT 38
         "   "   "   – RIDGE 52
         JACK RAFTER   – EDGE 34
         PURLIN   – EDGE 56
         "   – SIDE 53.5

JACK RAFTERS 333 mm **CENTRES DECREASE** 498   (in mm to 999 and
                 400 "   "   "   598   thereafter in m)
                 500 "   "   "   747
                 600 "   "   "   896

Run of Rafter	0.1	0.2	0.3	0.4	0.5	0.6	0.7	0.8	0.9	1.0
Length of Rafter	0.149	0.299	0.448	0.598	0.747	0.897	1.046	1.196	1.345	1.494
Length of Hip	0.18	0.36	0.539	0.719	0.899	1.079	1.259	1.438	1.618	1.798

Calculating length and cutting angles

RISE OF COMMON RAFTER 1.15 m **PER METRE OF RUN** **PITCH** 49°

BEVELS:	COMMON RAFTER	– SEAT	49
	" "	– RIDGE	41
	HIP OR VALLEY	– SEAT	39
	" " "	– RIDGE	51
	JACK RAFTER	– EDGE	33.5
	PURLIN	– EDGE	56.5
	"	– SIDE	53

JACK RAFTERS	333 mm	**CENTRES DECREASE**	508	(in mm to 999 and
	400 "	" "	610	thereafter in m)
	500 "	" "	762	
	600 "	" "	914	

Run of Rafter	0.1	0.2	0.3	0.4	0.5	0.6	0.7	0.8	0.9	1.0
Length of Rafter	0.152	0.305	0.457	0.61	0.762	0.915	1.067	1.219	1.372	1.524
Length of Hip	0.182	0.365	0.547	0.729	0.912	1.094	1.276	1.458	1.641	1.823

Goss's roofing ready reckoner

RISE OF COMMON RAFTER 1.192 m **PER METRE OF RUN**   **PITCH** 50°

BEVELS: COMMON RAFTER   – SEAT   50
         "   "   – RIDGE 40
         HIP OR VALLEY   – SEAT   40
         "   "   "   – RIDGE 50
         JACK RAFTER   – EDGE   32.5
         PURLIN   – EDGE   57.5
         "   – SIDE   52.5

JACK RAFTERS 333 mm **CENTRES DECREASE** 518 (in mm to 999 and
                 400 "   "   "   622   thereafter in m)
                 500 "   "   "   778
                 600 "   "   "   934

Run of Rafter	0.1	0.2	0.3	0.4	0.5	0.6	0.7	0.8	0.9	1.0
Length of Rafter	0.156	0.311	0.467	0.622	0.778	0.933	1.089	1.246	1.4	1.556
Length of Hip	0.185	0.37	0.555	0.74	0.925	1.11	1.295	1.48	1.664	1.849

Calculating length and cutting angles

RISE OF COMMON RAFTER 1.235 m **PER METRE OF RUN**   **PITCH** 51°

BEVELS: COMMON RAFTER – SEAT 51
            ″   ″   – RIDGE 39
         HIP OR VALLEY   – SEAT 41
            ″   ″   ″ – RIDGE 49
         JACK RAFTER   – EDGE 32
         PURLIN   – EDGE 58
            ″   – SIDE 52

JACK RAFTERS 333 mm **CENTRES DECREASE** 529   (in mm to 999 and
                  400 ″   ″   636   thereafter in m)
                  500 ″   ″   795
                  600 ″   ″   953

Run of Rafter	0.1	0.2	0.3	0.4	0.5	0.6	0.7	0.8	0.9	1.0
Length of Rafter	0.159	0.318	0.477	0.636	0.795	0.953	1.012	1.271	1.43	1.589
Length of Hip	0.188	0.375	0.563	0.751	0.936	1.126	1.314	1.502	1.69	1.877

Goss's roofing ready reckoner

RISE OF COMMON RAFTER 1.28 m **PER METRE OF RUN** **PITCH** 52°

BEVELS:	COMMON RAFTER	– SEAT	52
	" "	– RIDGE	38
	HIP OR VALLEY	– SEAT	42
	" " "	– RIDGE	48
	JACK RAFTER	– EDGE	31.5
	PURLIN	– EDGE	58.5
	"	– SIDE	52

JACK RAFTERS 333 mm **CENTRES DECREASE** 541 (in mm to 999 and
 400 " " " 650 thereafter in m)
 500 " " " 812
 600 " " " 974

Run of Rafter	0.1	0.2	0.3	0.4	0.5	0.6	0.7	0.8	0.9	1.0
Length of Rafter	0.162	0.325	0.487	0.65	0.812	0.974	1.137	1.299	1.462	1.624
Length of Hip	0.191	0.381	0.572	0.763	0.954	1.144	1.335	1.526	1.717	1.907

Calculating length and cutting angles

RISE OF COMMON RAFTER   1.327 m **PER METRE OF RUN**   **PITCH**   53°

BEVELS: COMMON RAFTER   – SEAT   53
                   ″   ″   – RIDGE   37
                HIP OR VALLEY   – SEAT   43
                 ″   ″   ″   – RIDGE   47
                JACK RAFTER   – EDGE   31
                PURLIN   – EDGE   59
                   ″   – SIDE   51.5

JACK RAFTERS   333 mm   **CENTRES DECREASE**   553   (in mm to 999 and
                       400 ″   ″   ″   665   thereafter in m)
                       500 ″   ″   ″   831
                       600 ″   ″   ″   997

Run of Rafter	0.1	0.2	0.3	0.4	0.5	0.6	0.7	0.8	0.9	1.0
Length of Rafter	0.166	0.332	0.498	0.665	0.831	0.997	1.163	1.329	1.495	1.662
Length of Hip	0.194	0.388	0.582	0.776	0.97	1.164	1.358	1.551	1.745	1.939

RISE OF COMMON RAFTER 1.376 m **PER METRE OF RUN** **PITCH** 54°

BEVELS: COMMON RAFTER – SEAT 54
            ″   ″   – RIDGE 36
         HIP OR VALLEY   – SEAT 44
            ″   ″   ″   – RIDGE 46
         JACK RAFTER   – EDGE 30.5
         PURLIN   – EDGE 59.5
            ″   – SIDE 51

JACK RAFTERS 333 mm **CENTRES DECREASE** 567 (in mm to 999 and
                  400 ″   ″   ″   680 thereafter in m)
                  500 ″   ″   ″   850
                  600 ″   ″   ″   1.021

Run of Rafter	0.1	0.2	0.3	0.4	0.5	0.6	0.7	0.8	0.9	1.0
Length of Rafter	0.17	0.34	0.51	0.681	0.851	1.021	1.191	1.361	1.531	1.701
Length of Hip	0.197	0.395	0.592	0.789	0.987	1.184	1.381	1.579	1.776	1.973

Calculating length and cutting angles

RISE OF COMMON RAFTER 1.428 m **PER METRE OF RUN** **PITCH** 55°

BEVELS:	COMMON RAFTER	– SEAT	55
	" "	– RIDGE	35
	HIP OR VALLEY	– SEAT	45.5
	" " "	– RIDGE	44.5
	JACK RAFTER	– EDGE	30
	PURLIN	– EDGE	60
	"	– SIDE	50.5

JACK RAFTERS 333 mm **CENTRES DECREASE** 580 (in mm to 999 and
             400 "   "   "   697   thereafter in m)
             500 "   "   "   872
             600 "   "   "   1.046

Run of Rafter	0.1	0.2	0.3	0.4	0.5	0.6	0.7	0.8	0.9	1.0
Length of Rafter	0.174	0.349	0.523	0.697	0.872	1.046	1.22	1.395	1.569	1.743
Length of Hip	0.201	0.402	0.603	0.804	1.005	1.206	1.407	1.608	1.809	2.01

Goss's roofing ready reckoner

RISE OF COMMON RAFTER 1.5 m **PER METRE OF RUN** **PITCH** 56° 18′
(Italian pitch)

BEVELS: COMMON RAFTER — SEAT 56.5
        ″   ″   — RIDGE 33.5
        HIP OR VALLEY — SEAT 46.5
        ″   ″   ″   — RIDGE 43.5
        JACK RAFTER   — EDGE 29
        PURLIN   — EDGE 61
        ″   — SIDE 50

JACK RAFTERS 333 mm **CENTRES DECREASE** 600 (in mm to 999 and
                 400 ″   ″   ″   720 thereafter in m)
                 500 ″   ″   ″   900
                 600 ″   ″   ″   1.082

Run of Rafter	0.1	0.2	0.3	0.4	0.5	0.6	0.7	0.8	0.9	1.0
Length of Rafter	0.18	0.361	0.541	0.721	0.902	1.082	1.262	1.442	1.623	1.803
Length of Hip	0.206	0.412	0.618	0.824	1.031	1.237	1.443	1.649	1.855	2.061

Calculating length and cutting angles

RISE OF COMMON RAFTER 1.6 m PER METRE OF RUN   PITCH 58°

BEVELS: COMMON RAFTER   – SEAT   58
                    ″   ″   – RIDGE 32
         HIP OR VALLEY   – SEAT   48.5
         ″   ″   ″   – RIDGE 41.5
         JACK RAFTER   – EDGE   28
         PURLIN   – EDGE   62
         ″   – SIDE   49.5

JACK RAFTERS 333 mm **CENTRES DECREASE** 628 (in mm to 999 and
                             400 ″   ″   ″   755 thereafter in m)
                             500 ″   ″   ″   944
                             600 ″   ″   ″   1.132

Run of Rafter	0.1	0.2	0.3	0.4	0.5	0.6	0.7	0.8	0.9	1.0
Length of Rafter	0.189	0.377	0.566	0.755	0.944	1.132	1.321	1.51	1.698	1.887
Length of Hip	0.214	0.428	0.641	0.855	1.069	1.283	1.496	1.71	1.924	2.136

Goss's roofing ready reckoner

RISE OF COMMON RAFTER 1.664 m **PER METRE OF RUN** **PITCH** 59°

BEVELS:	COMMON RAFTER	– SEAT	59
	″ ″	– RIDGE	31
	HIP OR VALLEY	– SEAT	49.5
	″ ″ ″	– RIDGE	40.5
	JACK RAFTER	– EDGE	27.5
	PURLIN	– EDGE	62.5
	″	– SIDE	49.5

JACK RAFTERS 333 mm **CENTRES DECREASE** 647   (in mm to 999 and
                  400 ″   ″   ″   777   thereafter in m)
                  500 ″   ″   ″   971
                  600 ″   ″   ″   1.165

Run of Rafter	0.1	0.2	0.3	0.4	0.5	0.6	0.7	0.8	0.9	1.0
Length of Rafter	0.194	0.398	0.582	0.777	0.971	1.165	1.359	1.553	1.747	1.942
Length of Hip	0.218	0.437	0.655	0.874	1.092	1.31	1.529	1.747	1.965	2.184

Calculating length and cutting angles

RISE OF COMMON RAFTER 1.732 m **PER METRE OF RUN** **PITCH** 60°
(Equilateral pitch)

BEVELS: COMMON RAFTER   – SEAT   60
         ″   ″   – RIDGE 30
         HIP OR VALLEY   – SEAT   51
         ″   ″   ″   – RIDGE 39
         JACK RAFTER   – EDGE   26.5
         PURLIN   – EDGE   63.5
         ″   – SIDE   49

JACK RAFTERS 333 mm **CENTRES DECREASE**   666   (in mm to 999 and
                  400 ″   ″   ″   800   thereafter in m)
                  500 ″   ″   ″   1.000
                  600 ″   ″   ″   1.200

Run of Rafter	0.1	0.2	0.3	0.4	0.5	0.6	0.7	0.8	0.9	1.0
Length of Rafter	0.2	0.4	0.6	0.8	1.0	1.2	1.4	1.6	1.8	2.0
Length of Hip	0.224	0.447	0.671	0.894	1.118	1.342	1.565	1.789	2.012	2.236

Goss's roofing ready reckoner

RISE OF COMMON RAFTER 1.804 m **PER METRE OF RUN**   **PITCH** 61°

BEVELS: COMMON RAFTER – SEAT 61
           "   "   – RIDGE 29
         HIP OR VALLEY – SEAT 52
           "   "   " – RIDGE 38
         JACK RAFTER   – EDGE 26
         PURLIN   – EDGE 64
           "   – SIDE 49

JACK RAFTERS 333 mm **CENTRES DECREASE** 687 (in mm to 999 and
                  400 "   "   "   825   thereafter in m)
                  500 "   "   "   1.032
                  600 "   "   "   1.238

Run of Rafter	0.1	0.2	0.3	0.4	0.5	0.6	0.7	0.8	0.9	1.0
Length of Rafter	0.206	0.413	0.619	0.825	1.031	1.238	1.444	1.65	1.857	2.063
Length of Hip	0.229	0.458	0.688	0.917	1.146	1.375	1.605	1.834	2.063	2.292

Calculating length and cutting angles

RISE OF COMMON RAFTER 1.88 m **PER METRE OF RUN**   **PITCH** 62°

BEVELS: COMMON RAFTER   – SEAT   62
               ″   ″   – RIDGE 28
         HIP OR VALLEY   – SEAT   53
           ″   ″   ″   – RIDGE 37
         JACK RAFTER   – EDGE   25
         PURLIN   – EDGE   65
           ″   – SIDE   48.5

JACK RAFTERS 333 mm **CENTRES DECREASE** 709   (in mm to 999 and
                 400 ″   ″   852   thereafter in m)
                 500 ″   ″   1.065
                 600 ″   ″   1.278

Run of Rafter	0.1	0.2	0.3	0.4	0.5	0.6	0.7	0.8	0.9	1.0
Length of Rafter	0.213	0.426	0.639	0.852	1.065	1.278	1.491	1.704	1.917	2.13
Length of Hip	0.235	0.471	0.706	0.941	1.177	1.412	1.647	1.882	2.118	2.353

Goss's roofing ready reckoner

RISE OF COMMON RAFTER 2.0 m PER METRE OF RUN **PITCH** 63° 26′
(Gothic pitch)

BEVELS:
COMMON RAFTER	– SEAT	63.5
″ ″	– RIDGE	26.5
HIP OR VALLEY	– SEAT	54.5
″ ″ ″	– RIDGE	35.5
JACK RAFTER	– EDGE	24
PURLIN	– EDGE	66
″	– SIDE	48

JACK RAFTERS 333 mm **CENTRES DECREASE** 745 (in mm to 999 and
                               400 ″   ″   ″   894   thereafter in m)
                               500 ″   ″   ″   1.118
                               600 ″   ″   ″   1.342

Run of Rafter	0.1	0.2	0.3	0.4	0.5	0.6	0.7	0.8	0.9	1.0
Length of Rafter	0.224	0.447	0.671	0.894	1.118	1.342	1.565	1.789	2.012	2.236
Length of Hip	0.245	0.49	0.735	0.98	1.225	1.47	1.715	1.96	2.205	2.45

Calculating length and cutting angles

RISE OF COMMON RAFTER 2.145 m **PER METRE OF RUN**   **PITCH** 65°

BEVELS:	COMMON RAFTER	– SEAT	65
	″ ″	– RIDGE	25
	HIP OR VALLEY	– SEAT	56.5
	″ ″ ″	– RIDGE	33.5
	JACK RAFTER	– EDGE	23
	PURLIN	– EDGE	67
	″	– SIDE	48

JACK RAFTERS 333 mm **CENTRES DECREASE** 788 (in mm to 999 and
   400 ″   ″   ″   946   thereafter in m)
   500 ″   ″   ″   1.183
   600 ″   ″   ″   1.420

Run of Rafter	0.1	0.2	0.3	0.4	0.5	0.6	0.7	0.8	0.9	1.0
Length of Rafter	0.237	0.473	0.71	0.946	1.183	1.42	1.656	1.893	2.13	2.366
Length of Hip	0.257	0.514	0.771	1.028	1.284	1.541	1.798	2.055	2.312	2.569

RISE OF COMMON RAFTER 2.246 m **PER METRE OF RUN**   **PITCH** 66°

BEVELS: COMMON RAFTER – SEAT 66
　　　　　　　　″　　　　″　 – RIDGE 24
　　　　　HIP OR VALLEY　 – SEAT 58
　　　　　″　　″　　″　　 – RIDGE 32
　　　　　JACK RAFTER　　 – EDGE 22
　　　　　PURLIN　　　　　– EDGE 68
　　　　　　　″　　　　　 – SIDE 47.5

JACK RAFTERS　333 mm **CENTRES DECREASE**　819　(in mm to 999 and
　　　　　　　　 400 ″　　　　″　　　　　″　　984　thereafter in m)
　　　　　　　　 500 ″　　　　″　　　　　″　 1.230
　　　　　　　　 600 ″　　　　″　　　　　″　 1.475

Run of Rafter	0.1	0.2	0.3	0.4	0.5	0.6	0.7	0.8	0.9	1.0
Length of Rafter	0.246	0.492	0.738	0.983	1.229	1.475	1.721	1.967	2.213	2.459
Length of Hip	0.265	0.531	0.796	1.062	1.327	1.593	1.858	2.123	2.389	2.654

Calculating length and cutting angles

RISE OF COMMON RAFTER 2.356 m **PER METRE OF RUN**   **PITCH** 67°

BEVELS:	COMMON RAFTER	– SEAT	67
	" "	– RIDGE	23
	HIP OR VALLEY	– SEAT	59
	" " "	– RIDGE	31
	JACK RAFTER	– EDGE	21.5
	PURLIN	– EDGE	68.5
	"	– SIDE	47.5

JACK RAFTERS 333 mm **CENTRES DECREASE** 852 (in mm to 999 and thereafter in m)
  400 " " " 1.024
  500 " " " 1.280
  600 " " " 1.535

Run of Rafter	0.1	0.2	0.3	0.4	0.5	0.6	0.7	0.8	0.9	1.0
Length of Rafter	0.256	0.512	0.768	1.024	1.28	1.536	1.792	2.047	2.303	2.559
Length of Hip	0.275	0.55	0.824	1.099	1.374	1.649	1.922	2.198	2.473	2.748

Goss's roofing ready reckoner

RISE OF COMMON RAFTER 2.475 m **PER METRE OF RUN**   **PITCH** 68°

BEVELS:
- COMMON RAFTER — SEAT 68
- " " — RIDGE 22
- HIP OR VALLEY — SEAT 60.5
- " " " — RIDGE 29.5
- JACK RAFTER — EDGE 20.5
- PURLIN — EDGE 69.5
- " — SIDE 47

JACK RAFTERS 333 mm **CENTRES DECREASE** 889 (in mm to 999 and thereafter in m)
400 " " " 1.068
500 " " " 1.335
600 " " " 1.601

Run of Rafter	0.1	0.2	0.3	0.4	0.5	0.6	0.7	0.8	0.9	1.0
Length of Rafter	0.267	0.534	0.801	1.068	1.335	1.602	1.869	2.136	2.403	2.669
Length of Hip	0.285	0.57	0.855	1.14	1.425	1.71	1.995	2.28	2.566	2.851

Calculating length and cutting angles

RISE OF COMMON RAFTER 2.605 m **PER METRE OF RUN**   **PITCH** 69°

	BEVELS:	COMMON RAFTER	– SEAT	69
		" "	– RIDGE	21
		HIP OR VALLEY	– SEAT	61.5
		" " "	– RIDGE	28.5
		JACK RAFTER	– EDGE	19.5
		PURLIN	– EDGE	70.5
		"	– SIDE	47

JACK RAFTERS 333 mm **CENTRES DECREASE** 930 (in mm to 999 and
                 400 "   "   "   1.116 thereafter in m)
                 500 "   "   "   1.395
                 600 "   "   "   1.674

Run of Rafter	0.1	0.2	0.3	0.4	0.5	0.6	0.7	0.8	0.9	1.0
Length of Rafter	0.279	0.558	0.837	1.116	1.395	1.674	1.953	2.232	2.511	2.79
Length of Hip	0.296	0.593	0.889	1.186	1.482	1.779	2.075	2.371	2.688	2.964

RISE OF COMMON RAFTER 2.747 m **PER METRE OF RUN**   **PITCH** 70°

BEVELS: COMMON RAFTER – SEAT   70
         "   "   – RIDGE 20
         HIP OR VALLEY   – SEAT   63
         "   "   "   – RIDGE 27
         JACK RAFTER   – EDGE   19
         PURLIN   – EDGE   71
         "   – SIDE   47

JACK RAFTERS 333 mm **CENTRES DECREASE**   975   (in mm to 999 and
                  400 "   "   "   1.170   thereafter in m)
                  500 "   "   "   1.462
                  600 "   "   "   1.754

Run of Rafter	0.1	0.2	0.3	0.4	0.5	0.6	0.7	0.8	0.9	1.0
Length of Rafter	0.292	0.585	0.877	1.17	1.462	1.754	2.047	2.339	2.631	2.924
Length of Hip	0.309	0.618	0.927	1.236	1.545	1.854	2.163	2.472	2.781	3.09

Calculating length and cutting angles

RISE OF COMMON RAFTER 2.904 m **PER METRE OF RUN**   **PITCH** 71°

BEVELS: COMMON RAFTER   – SEAT   71
                                                       ″   ″   – RIDGE 19
                        HIP OR VALLEY   – SEAT   64
                             ″   ″   ″   – RIDGE 26
                        JACK RAFTER   – EDGE   18
                        PURLIN   – EDGE   72
                             ″   – SIDE   46.5

JACK RAFTERS 333 mm **CENTRES DECREASE** 1.024   (in mm to 999 and
                                   400 ″   ″   ″   1.229   thereafter in m)
                                   500 ″   ″   ″   1.536
                                   600 ″   ″   ″   1.843

Run of Rafter	0.1	0.2	0.3	0.4	0.5	0.6	0.7	0.8	0.9	1.0
Length of Rafter	0.307	0.614	0.921	1.229	1.536	1.843	2.15	2.457	2.764	3.072
Length of Hip	0.323	0.646	0.969	1.292	1.615	1.938	2.261	2.584	2.907	3.23

Goss's roofing ready reckoner

RISE OF COMMON RAFTER 3.078 m **PER METRE OF RUN**   **PITCH** 72°

	BEVELS:	COMMON RAFTER	– SEAT	72
		″ ″	– RIDGE	18
		HIP OR VALLEY	– SEAT	65.5
		″ ″ ″	– RIDGE	24.5
		JACK RAFTER	– EDGE	17
		PURLIN	– EDGE	73
		″	– SIDE	46.5

JACK RAFTERS 333 mm **CENTRES DECREASE** 1.078 (in mm to 999 and
   400 ″   ″   ″   1.294   thereafter in m)
   500 ″   ″   ″   1.618
   600 ″   ″   ″   1.942

Run of Rafter	0.1	0.2	0.3	0.4	0.5	0.6	0.7	0.8	0.9	1.0
Length of Rafter	0.324	0.647	0.971	1.294	1.618	1.942	2.266	2.589	2.912	3.236
Length of Hip	0.339	0.677	1.016	1.355	1.694	2.032	2.371	2.71	3.048	3.387

Calculating length and cutting angles

RISE OF COMMON RAFTER 3.271 m **PER METRE OF RUN** **PITCH** 73°

	BEVELS:	COMMON RAFTER	– SEAT	73
		" "	– RIDGE	17
		HIP OR VALLEY	– SEAT	66.5
		" " "	– RIDGE	23.5
		JACK RAFTER	– EDGE	16.5
		PURLIN	– EDGE	73.5
		"	– SIDE	46.5

JACK RAFTERS 333 mm **CENTRES DECREASE** 1.140 (in mm to 999 and
                 400 "   "   "   1.368 thereafter in m)
                 500 "   "   "   1.710
                 600 "   "   "   2.052

Run of Rafter	0.1	0.2	0.3	0.4	0.5	0.6	0.7	0.8	0.9	1.0
Length of Rafter	0.342	0.684	1.026	1.368	1.71	2.052	2.394	2.736	3.078	3.42
Length of Hip	0.356	0.713	1.069	1.425	1.782	2.138	2.494	2.851	3.207	3.563

Goss's roofing ready reckoner

RISE OF COMMON RAFTER 3.487 m **PER METRE OF RUN** **PITCH** 74°

BEVELS: COMMON RAFTER – SEAT 74
　　　　　　　　 ″　　　　″　　– RIDGE 16
　　　　　HIP OR VALLEY – SEAT 68
　　　　　　″　　″　　　″　　– RIDGE 22
　　　　　JACK RAFTER – EDGE 15.5
　　　　　PURLIN – EDGE 74.5
　　　　　　″　　　　　　– SIDE 46

JACK RAFTERS 333 mm **CENTRES DECREASE** 1.209 (in mm to 999 and
　　　　　　　　　400 ″　　　　″　　　　　　″　　1.451　thereafter in m)
　　　　　　　　　500 ″　　　　″　　　　　　″　　1.814
　　　　　　　　　600 ″　　　　″　　　　　　″　　2.177

Run of Rafter	0.1	0.2	0.3	0.4	0.5	0.6	0.7	0.8	0.9	1.0
Length of Rafter	0.363	0.726	1.088	1.451	1.814	2.177	2.54	2.902	3.265	3.628
Length of Hip	0.376	0.753	1.129	1.505	1.882	2.258	2.634	3.011	3.387	3.763

Calculating length and cutting angles

RISE OF COMMON RAFTER 3.732 m **PER METRE OF RUN**   **PITCH** 75°

BEVELS: COMMON RAFTER – SEAT 75
          "   "   – RIDGE 15
         HIP OR VALLEY   – SEAT 69
          "   "   "   – RIDGE 21
         JACK RAFTER   – EDGE 14.5
         PURLIN   – EDGE 75.5
          "   – SIDE 46

JACK RAFTERS 333 mm **CENTRES DECREASE** 1.289 (in mm to 999 and
                 400 "   "   "   1.546   thereafter in m)
                 500 "   "   "   1.932
                 600 "   "   "   2.318

Run of Rafter	0.1	0.2	0.3	0.4	0.5	0.6	0.7	0.8	0.9	1.0
Length of Rafter	0.386	0.773	1.159	1.545	1.932	2.318	2.705	3.091	3.477	3.864
Length of Hip	0.399	0.789	1.197	1.596	1.996	2.395	2.794	3.193	3.592	3.991

5 WALL PLATES – STRAPPING AND GABLE STRAPPING

Earlier in this book it was stated that the wall plate is the foundation to the roof, and like all foundations needs to be sound and secure. The wall plate should be strapped down to the building structure; this is usually done by using steel straps as illustrated in Figure 5.1(a). The straps can either be built into the brick wall below, or face-fixed by screwing into the wall. On timber framed housing the plate may be adequately secured to the frame by nailing at centres specified by the designer.

Gable walls, especially those on steep pitched roof constructions where the gable is very tall, rely on the roof for their stability and NOT the other way around. Wind blowing on one gable exerts a pressure on it pushing it into the roof, whilst at the opposite end it creates a suction which attempts to suck the gable from the roof. It is therefore a requirement of the building regulations that the gables must be adequately tied back into the roof to give them support. Figure 5.1(b) illustrates a typical gable end restraint system on a trussed rafter roof, but this equally applies to traditional roof construction. Building the purlins into the gable will help, but will only be effective if the purlin is mechanically fixed to the wall with some additional form of cleat or strap. Straps are normally placed at approximately 2 m centres, and it is essential that the strap is supported by solid blocking beneath it to ensure that it does not buckle, and that the last rafter is solidly blocked to the gable wall itself.

Wall plates – strapping and gable strapping

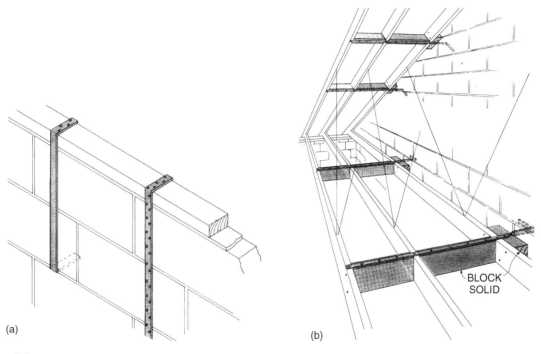

(a)

(b)

Figure 5.1

6 WIND BRACING AND OPENINGS FOR DORMERS AND ROOF WINDOWS

From the previous chapter it can be seen that wind plays a considerable part in destabilising a structure and measures must be taken to ensure the stability of the roof construction in high wind situations. From the previous chapter strapping the gables to the roof has ensured their integrity with the roof, but apart from the binders, purlins and ridge, which connect the rafter members on a horizontal plane, the roof structure is still no more than a number of vertical members of timber connected with a limited number of nails to the members mentioned above. The roof is in effect no more than a set of dominoes standing vertically on their ends and which can be easily made to fall if a pressure is applied to the last one in line; in the case of a roof this would be the gable end. To prevent the domino toppling effect, wind bracing is introduced into the structure to triangulate it on a vertical plane. The elements being discussed here are set out in Figure 6.1, which details the bracing required for a typical trussed rafter roof, much of this being applicable to a traditionally constructed roof without hips. Even on a hip roof where the ridge is twice as long as the length on plan of the hip itself it is wise to introduce wind bracing.

Wind bracing is usually timber typically of 25 × 100 mm in cross section, fitted from wall plate to the ridge at an angle of approximately 45° on the underside of the rafters. At each rafter crossing the wind bracing should be nailed to the rafter with 3 nails. The braces should be fitted from the foot of the gable to the ridge on both sides of the roof and then at 45° back down again to the plate over the entire length of the roof. This is brace F in Figure 6.1; brace H may well be replaced by a purlin in a traditionally cut

Wind bracing and openings for windows

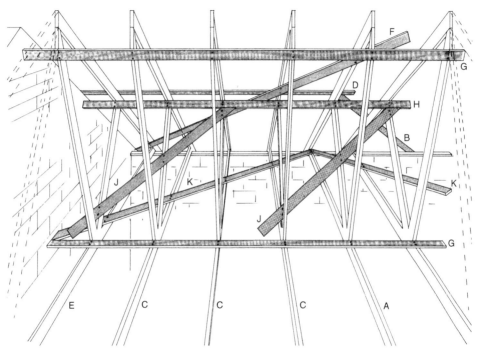

Figure 6.1 Bracing required for typical trussed rafter roof.

roof, and brace G by the ridge board. Brace G at ceiling joist level, may well be ceiling joist binders on a traditionally cut roof, whilst K should be fitted to all types of roof construction. Brace J applies only to trussed rafter roof construction.

OPENINGS FOR DORMERS AND ROOF WINDOWS

If part or the whole of the roof is to be used as a living space, or simply a hobby room, or perhaps some natural light is required in the roof void for some other purpose, then openings may have to be formed in the roof structure. The following text gives some guidance on openings in newly built traditional cut roofs, conversion work in traditional cut roofs, new trussed rafter roofs, and existing trussed rafter roofs.

Traditional Cut Roofs (not Trussed Rafter Construction)

New Build Roofs
Openings for dormers and roof windows may be formed by adding timbers to the sides of the opening as set out in Figures 6.2(a) and (b) taken from the book *Roof Construction and Loft Conversion*, by this author. It is strongly recommended that all timber connections, i.e. trimmed rafter to trimmer and trimmer to double or triple rafter, are made with metal connection plates, fixed using the correct nails or screws, as this gives a stronger joint than simply attempting to angular or 'tosh' nail. Whilst 'tosh' nailing is a traditional way of connecting roof timbers, it is very prone to splitting the timber, resulting in a poor joint.

Wind bracing and openings for windows

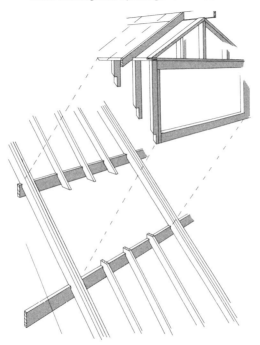

Figure 6.2 (a) Trimmed opening for dormer. For trimmer numbers required see Figure 6.2(b). Reproduced from C. N. Mindham (2006) *Roof Construction and Loft Conversion*, 4th edn. UK: Blackwell, p. 49.

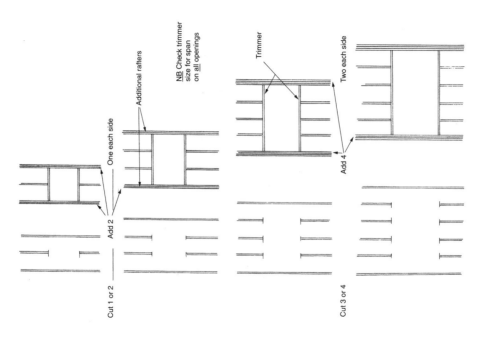

Figure 6.2 (b) Trimming construction rules. Reproduced from C. N. Mindham (2006) *Roof Construction and Loft Conversion*, 4th edn. UK: Blackwell, p. 50.

Conversion Work
The same rules as set out above will apply, but additionally a purlin may have to be removed to give adequate opening for the new dormer or roof window. In this case a further purlin should be inserted above and below the opening and securely fixed to its supporting walls or purlinpost, and the existing roof fitted to it BEFORE removing the old purlin and cutting the rafters away to form the roof opening. Additional rafters either side of the opening may not be required if the purlins are immediately above and below the opening itself. Trimmers should still be fitted to pick up the ends of the rafters to completely frame the opening. Possible roof window installations are shown in Figure 6.3.

Trussed Rafter Roofs

New Roofs
The trussed rafter roof designer will take any openings in the roof, including those for chimneys, into account in the calculation of the whole roof design. Specific rules apply to forming openings in trussed rafter roofs and on NO account should the rules given above for traditional roofs be applied to trussed rafter constructions. Trussed rafter roofs are generally now a 'whole roof' design, and not simply the supply of trussed rafters to be erected as the builder wishes. Make use of this design service available through your trussed rafter supplier to ensure a sound roof structure.

Existing Trussed Rafter Roofs
Trussed rafters should NEVER be cut, as this will endanger the whole roof structure. Consult the original roof designer or manufacturer if known and seek their advice on any modifications to form openings in the roof that may be proposed. Details of the original manufacturer should be found on a card fixed under the nail plate at the ridge of the trussed rafter; this will give details of the manufacturer as well

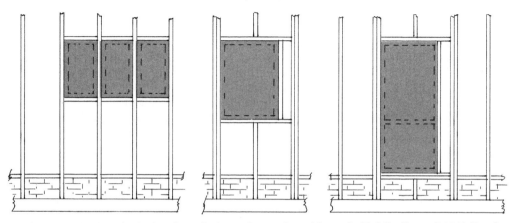

Figure 6.3 Some typical roof window installations. Reproduced from C. N. Mindham (2006) *Roof Construction and Loft Conversion*, 4th edn. UK: Blackwell, p. 216.

as the nail plate system provider. The identity card was not until fairly recently standardised as being fixed at the ridge, and it may therefore be fixed under any other plate on the structure. Early trussed rafters will have no form of identification of their manufacturer. In that case there is no alternative but to consult a structural engineer.

However, as most trussed rafter roofs have trussed rafters spaced at 600 mm centres, it is possible to buy roof windows that are designed specifically to fit neatly between the trussed rafters, thus avoiding any structural modifications to the roof structure. It is possible to couple a number of these windows side by side along the length of the roof, giving the effect of a much larger single opening. It clearly avoids the expense of altering the roof structure, and is much more simple and cost effective. As this approach results in a multiple number of small windows, it affords more ventilation options than one large window. The left hand illustration in Figure 6.3 indicates multiple roof window installation in a trussed rafter or indeed in a traditionally constructed cut roof.

7 ROOFING METALWORK AND FIXINGS

The use of smaller timber sections, particularly in engineered roofs such as trussed rafter prefabricated assemblies, has led to the increased use of metalwork to join the various members together. Traditional 'tosh' or 'skew' nailing can easily split these smaller timber sections, resulting in a poor connection. Figure 7.1 illustrates some items of roofing metalwork.

All metalwork used in roof construction should be galvanised; this includes the nails, to prevent corrosion. The modern roof can suffer from localised condensation, especially if not adequately ventilated and this could lead to premature corrosion of nails and fixings. Whilst a galvanised nail has a slightly rough surface compared to a plain wire nail and therefore gives an improved resistance to movement in the joint, many of the metal to timber connections are specified to be fixed with square twisted galvanised nails which give a far improved performance in the joint.

Typical roofing metalwork would be as follows:

(a) Wall plate straps cross section 30 mm × 1.5/2 mm
(b) Gable restraint straps 30 mm × 5 mm
(c) Trussed rafter clips to hold truss or rafter to wall plate
(d) Hip corner tie to hold hip to wall plate at corner
(e) Girder truss shoes to carry trussed rafters on girder truss support points
(f) Multi nail plates for coupling timbers into longer lengths
(g) Framing anchors to connect various trimmed openings and elements in the ceiling joist structure

Roofing metalwork and fixings

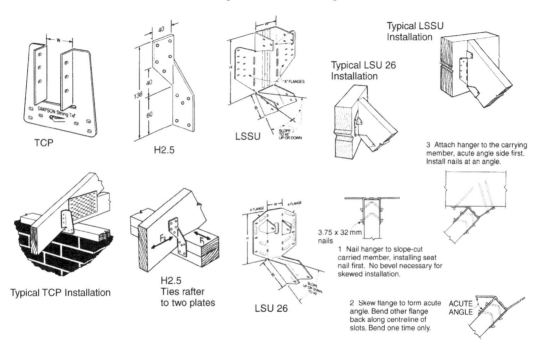

Figure 7.1 Some items of roofing metalwork.

Care must be taken to use the nails specified by the metalwork manufacturer: both in the type of nail to be used and the number of nails to be used in each joint.

NAILS, BOLTS AND SCREWS

Whilst traditionally roof structures have been connected using nails and continue to be so, there is an ever increasing use of light metal connectors as illustrated in Figure 7.1, complete with their specialist nails. The use of traditional large round wire nails (galvanised for roof structures), can lead to serious splitting of the timbers, thus resulting in a poor strength of joint. It is also a practised skill to be able to nail a rafter to a wall plate using 'tosh' or 'skew' nailing, and often leads to misplacement of the rafter itself. Again, a better method is to use a framing anchor or a truss clip as illustrated. However, with the ready availability of powerful battery operated drill/drivers, the application of bolts and high tensile extremely long steel screws makes the construction of a roof not only easier, but more precise, and definitely stronger. If bolts are to be used, then either a hexagon head or a cup square head coach bolt could be used, but in either case a large square or round washer should be placed under both the head of the bolt and the nut to avoid crushing the timber. Washers are traditional 50 mm × 50 mm × 3 mm thick or 50 mm or 63 mm diameter. DO NOT OVER TIGHTEN, as this does not necessarily greatly increase the strength of the connection; the object is not to distort the washer by bedding it into the timber, as this could cause splitting.

The modern especially designed long timber screws are manufactured from carbon steel and are epoxy coated to provide corrosion protection while at the same time lowering friction when driving the screw. These relatively slender screws do not require pilot holes, and are driven by a special adaptor tool provided with the screws, driven by a minimum of 18 V battery powered drill driver. The application is extremely fast and provides a very strong joint. These screws are particularly efficient and time saving compared to a bolt when making connections between two trimmers such as those at the side of roof window openings. Their lengths range from 66 to 135 mm. One manufacturer is Fasten Master and these are available from many builders' merchants and timber suppliers. For further details see Bibliography.

8 ROOF COVERINGS – FELT, BATTENS AND TILES

UNDERLAY

Underlay is a thin sheet material laid over the top of the rafters, held in place by the roof covering supports or battens. There are principally two types of underlay: vapour permeable and vapour impermeable. The permeable type allows water vapour rising to the roof space to escape through the underlay, whilst not allowing any rain that may penetrate the roof covering to enter the building. The impermeable type is water and vapour proof in both directions, and can therefore give rise to moisture from water vapour being trapped within the roof space void. Ventilation of the roof in this case is essential – more information on this matter later.

Traditionally, roof underlays were a bitumen sheet reinforced with hessian type cloth. This is still used but clearly falls into the impermeable type. The more modern underlays conform to the vapour permeable standard. The decision on which to use will be influenced by the intended use of the roof void, i.e. storage or living loft, and if a 'cold' or 'warm' roof is being designed – see more information on this aspect under Roof Void Ventilation below.

Unless the roof structure is situated in Scotland, it is unlikely to be covered with a solid boarding or sarking. If it is, then the roof underlay will be only partially supported by the sarking because counter battening should be used to provide ventilation between the underlay and the sarking itself. For the majority of the UK, no sarking will be used and the underlay will be unsupported. Care must therefore be taken to avoid too much droop between the rafters, and particularly to avoid droop at the eaves behind the fascia. If this occurs, and rain penetrates the roof tiles, then ponding will occur behind the fascia, leading possibly to water penetrating adjacent to the wall plates and into the soffit area. An underlay support in some form of exterior boarding should be placed immediately behind the fascia to support the underlay. At any point on the roof, great care should be taken not to damage the underlay whilst battens and tiles are being laid on the roof. See Figure 8.1 for underlay detailing.

The purpose of the underlay is to prevent any water that may penetrate the roof covering entering the building, by draining it away into the gutter. It must therefore be applied by a method to ensure that the roof cannot be penetrated. As all underlays are produced in rolls, both vertical and horizontal joints in the material are unavoidable (except on roofs where the roll length exceeds the roof length between gables), and guidance on horizontal laps is given in Figure 8.2. This information is for not fully supported underlays.

BATTENS

Battens support the roof covering of tiles, slates or shingles, spanning from rafter to rafter and provide a medium into which the tile fixing nails can be bedded. They also provide a 'ladder' for ease of walking on the roof, but care must be taken to stay along the rafter line and not to accidentally penetrate the underlay. Any damage to the underlay must be repaired in such a way as to avoid water penetration – sticking a piece on the surface will probably not be adequate.

The Following Aspects Must be Considered When Specifying Battens
(a) Weight of the tiles or roof covering
(b) Method of fixing required by the tiles, i.e. a nib depth and nailing recommendation
(c) Distance between the rafters, i.e. span of the battens
(d) Durability of the material of the batten
(e) Strength of the batten

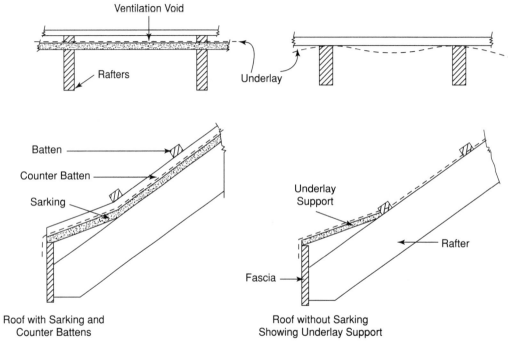

Figure 8.1 Underlay detailing.

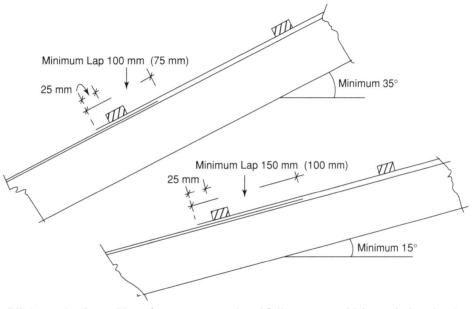

Figure 8.2 Minimum horizontal laps for unsupported and fully supported (shown in brackets) underlay.

In general, timber battens will be used as they are economical, readily available, easily handled, easy to nail to and easy to cut. They are, however, generally of a small cross sectional area, typically 38 mm wide × 25 mm deep or 50 mm wide × 25 mm deep. Therefore, they should be of good quality timber. Any defects, splits, knots or wane will greatly affect the strength of the batten and, as the roofer may end up standing on the batten between rafters, they must be considered as the rung of a ladder. The British Standard indeed requires that the battens are free of decay, insect attack, splits, shakes, knots or knot holes greater than one-third the width of the batten. They should be a minimum of 1200 mm in length and have both ends supported.

The size of the battens mentioned above should be checked with the technical literature of the roof covering tile or slate manufacturer, but in general terms 38 mm × 25 mm will span rafters of 450 mm centres, and 50 mm × 25 mm will span between 450 and 600 mm centres. If the roof covering is exceptionally heavy, such as stone slates as distinct from thin natural Welsh slate, then guidance should be sought from the provider or from an engineer.

Timber battens should be treated with preservative and are generally seen as requiring a minimum 60 year service life. The latest British Standard assigns a hazard class to building timbers, depending on their exposure to fungal and insect attack. Tiling battens are classified as hazard class 2. To place this in perspective, roof structural members are generally in hazard class 1, whilst members exposed directly to wetting, i.e. the fascia, the barge and other exposed timbers such as exposed rafter feet, are in hazard class 3 therefore needing greater protection. By specifying the hazard class, the preservative treatment processing plant will be able to advise on the correct preservative system to be used.

All preservatives contain chemicals to do their job and some can prove corrosive to certain metal fixings. It is therefore essential to ensure that nails used to fix battens to rafters, and nails used to fix tiles to battens are compatible with the preservatives used to treat the battens. Bright wire nails are NOT acceptable, galvanised nails may be, but check with the preservative supplier. Aluminium nails are commonly used, as are copper nails, especially for fixing tiles to battens. The tile manufacturer's recommendations should be sought to ensure nail (or screw) compatibility with tiles or slates specified.

INSULATION AND VENTILATION

The drive to reduce carbon emissions and therefore global warming is causing the government to set increasingly higher standards of thermal insulation for all buildings including dwellings. Part L of the *Building Regulations* sets out the requirements and has become a complex document, giving the standards to be achieved for the whole building (not just the roof or the walls, or any other element) for all new dwellings and refurbishment. However, different standards are set for domestic new and refurbishment structures.

It is not intended in this text to describe the calculations as they involve the 'whole' building, and this book is concerned only with the roof structure. Guidance on the insulation of the roof structures is set out below. The largest area of heat loss in a building, assuming normal brick and block or brick and timber frame construction, is that through the roof and it is for that reason that the standards indicate that the roof should offer the greatest resistance to heat loss. These standards are expressed as U values. To give an idea of the relative performance of the building element, the higher the value indicating the greater the heat transmission or loss, we have the following U values:

External walls 0.55 W/m^{2}K
Windows and doors 2 W/m^{2}K
Floors 0.25 W/m^{2}K
Roofs 0.16 W/m^{2}K

Part L of the *Building Regulations* deals with all aspects of the conservation of fuel and power. Whilst this concerns itself with the thermal performance of individual elements, it also recognises the effect on heat conservation caused by air changes within the building resulting from poor draught sealing; not only of the opening parts of windows and doors, but also any gaps which may be around the frames.

It can be seen, therefore, from the above that the thermal insulation performance requirement of the roof is over 12 times that of windows and doors. Paying great attention to the correct specification and installation of the insulation in the roof will have a major effect on the thermal efficiency and therefore the cost of heating the building. This is particularly true of refurbishment and extension projects where the original insulation may be negligible.

Warning!

Do NOT specify the insulation without careful consideration to the possibility of condensation forming in the roof space – this can lead to wet insulation and wet insulation DOES NOT offer any thermal resistance; it MUST be dry – see Roof Void Ventilation section below.

There are numerous types of insulation material available, all having their advantages and disadvantages in their ease of installation in a building and also, of course, their cost. Some are sold on their 'green' merits, being natural products and others are man made, consuming energy to produce them. One product, for instance, made of sheep's wool uses only 14% of the energy in its manufacture compared to that taken to produce glass fibre insulation. An indication of the insulation types available is set out below, but it must be stressed that this is by no means a complete list.

Rockwool, mineral wool, glass fibre

A traditional and readily available product in roll or slab (bat) form, easily handled, but some personal protection may be required in the form of gloves and masks.

Hi tech multilayer reflective insulation

A high performance for a relatively thin layer, easily applied in roll form, that reduces the risk of interstitial condensation (i.e. condensation forming within the insulation layer) because the surface layers are impermeable.

Polystyrene sheet
Lightweight rigid sheet particularly suitable for fixing between rafters in attic type construction – check for fire resistant standard.

Spray applied polyurethane foam
A product usually applied by specialist contractors, but available as a DIY spray-on foam. This type is particularly suitable for increasing the insulation in refurbishment works where rafters are not of regular shape, and this product also stops air leaks and air movements, but of course is vapour impermeable.

Rigid polyurethane sheet
Available in various thicknesses as a rigid sheet; applications similar to polystyrene sheet.

Blown fibre products
Insulation is by inert fibres usually applied by specialist contractors, with the fibres being blown into position. This generally means that it is suitable for ceilings and not for sloping surfaces.

Fleece
Natural sheep's wool especially prepared for insulation. Available in rigid bat form.

Fibre insulation board
Rigid low density fibre board, useful for attic applications between rafters. Available in various thicknesses.

To give an indication of the different thicknesses of insulation material required to meet the basic building regulation requirement for roof structures, 270 mm of rockwool would be needed but only 1 layer of approximately 25 mm of hi tech insulation plus 120 mm of mineral wool or 100 mm of polyurethane rigid sheet board. Whilst the rockwool insulation would probably be suitable for a roof ceiling insulation layer without loft space storage (there is no point in compressing the 270 mm by laying boards over the roof joists, as this defeats the insulation performance), the thinner materials (such as the polyurethane rigid sheet) will be more suitable for between rafter insulation in an attic or loft structure. The cost as laid must also be taken into consideration, and whilst the basic cost of the insulation may be relatively economic, if there is a high wastage factor then this could affect the 'as-laid' cost of the insulation. For further information from various manufacturers of some of the above products, see Bibliography.

Roof Void Ventilation

Why Ventilate?
Many old houses pre early 1900s had little or no insulation and indeed no underlay under the tiles or slates. Warm moist (vapour laden) air rose through the building into the roof void and was vented out through the numerous small gaps between the tiles. The roof timbers where thus protected from moisture build up within the roof void (assuming that the roof covering was kept in good condition), and lasted for centuries. The gaps did of course let in insects, some of which attacked the timbers, but most wood worm, which is the most common form of attack throughout the majority of the country, came from infestation in cheap plywoods brought in during the early 1900s in furniture.

Ever increasing standards to conserve energy result in higher levels of thermal insulation, often applied at ceiling level, which gives rise to a cold roof void. Vapour still rises from the house but now passes through the insulation and readily condenses into moisture on any cold surface within the roof space. This could simply be nails, metal connectors, and steel support systems that may be in the roof, and on trussed rafter roofs, the truss connector plates themselves. If there were still no underfelt within the roof space, the majority of this moisture would be ventilated and would cause no problem.

However, as there is often a bituminous felt underlay so common over the past several decades, this tends to keep the vapour within the building and can lead to formation of so much condensation that it can drip from the cold surfaces onto the ceiling below. This can in extreme cases lead the occupier to think that the roof is leaking. The author has inspected roof spaces where this condition is so acute that frost has formed on the underside of the felt in the roof space and when this thaws the amount of water descending on the ceiling can be significant. It is therefore necessary to ventilate this vapour from the roof void to keep the structure dry. If the project is a new construction, then building regulations set out rules for this ventilation, but it would be wise to follow these rules and provide ventilation in any refurbishment project even if that building is not subject to building regulations. This is simply good building practice.

Comparison of Warm Roofs and Cold Roofs
To better make provision for this ventilation, it must first be established whether the design is going to lead to a warm roof structure or a cold roof structure. A cold roof is that described above, i.e. insulation at ceiling level giving a cold roof void. A warm roof void is that where the insulation is provided at rafter level, i.e. as in an attic room, although it should be appreciated that there can be cold voids in an attic roof structure if there are spaces outside the habitable rooms. For instance, this could be the triangle which often occurs between the side wall and the eaves, and the small triangle which occurs above the ceiling up to the ridge. See Figure 8.3 for warm and cold roofs.

Controlling Moisture Laden Air Within the Structure
In the early 1900s houses described above with no underlay, the rafter level could be described as 'permeable', i.e. water vapour was free to escape through the gaps in the tiles. If, however, a bitumen impregnated cloth based traditional roofing underlay is used, this is non-vapour permeable, i.e. it will not let vapour pass through. There is now, however, a new generation of vapour permeable underlays

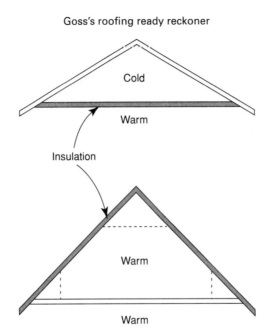

Figure 8.3 Cold and warm roof voids.

that will allow vapours to pass through but not allow water, which may penetrate the tiles into the building. There are standards set out in building regulations for the permeability of these underlays, and these should be checked against the product to be specified. In this case the roof void may not need ventilating.

A vapour permeable underlay works like ducks' feathers, in that it keeps out water from wind driven rain which may pass through the tiles, but allows the 'body' of the house to breathe by exhausting vapour. If this concept is difficult to appreciate, then the reader should consider the effects of hard exercise in a waterproof non-breathable jacket: our perspiration condenses on the inside of the material to make it wet inside, even when there is no wetting from the outside. This is exactly the same effect as the moisture within the house from bathing and cooking on the underside of an impermeable underlay. The new modern vapour permeable underlays can be compared with the latest high tech fabrics used on high quality waterproof jackets, which allow our perspiration to breathe out without making the jacket wet inside, but do not let water in.

How do we Decide Which Roof Type to Specify?
Do we want a cold or a warm roof?
Do we need the roof void for storage?
    Yes – design a warm roof
    No – design a cold roof.
Do we need to live in the attic space?
    Yes – design a warm roof
    No – design a cold roof.

Figure 8.4 illustrates various designs for cold and warm roof structures.

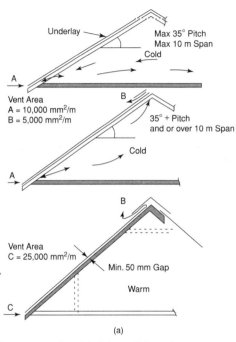

Figure 8.4 Designs for cold and warm roofs with (a) traditional non-permeable and (b) vapour permeable underlay. (c) Designs for mono pitch cold roof and attic warm roof with permeable underlay.

Felt, battens and tiles

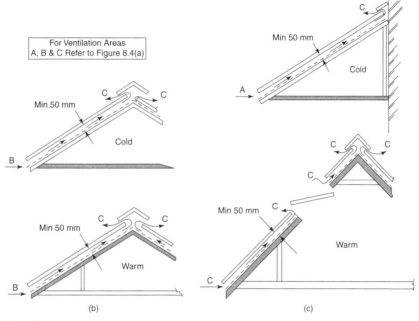

Figure 8.4 Continued.

In Scotland and Northern Ireland where fully boarded roofs are common, i.e. boards on top of the rafters or 'sarking' is used, ALL underlays should be treated as impermeable and ventilation provided accordingly. This construction is rarely used elsewhere in the UK, but if it is, the same rules apply as stated above. Refer to the *Building Regulations* for Scotland (Part C) and for Northern Ireland (Part B) for full details.

Economic Considerations

Heating Costs
Clearly a roof insulated at ceiling level gives the least heated volume in the property, but full ventilation of the cold roof void and good insulation of the water storage tanks and all roof void water pipework is essential. If the roof void is to be used for a limited storage, one has to consider if the stored goods would suffer from the thermal extremes suffered within a roof void between the heat of the summer and the extreme cold of the winter. Also, even in extreme conditions with ventilation, stored metal objects may attract some temporary condensation in cold conditions.

A warm roof, however, results in a larger heated volume which has a proportional effect on cost, and perhaps with limited rafter depth, it may prove more difficult to provide the required insulation. To gain this additional depth of rafter, it may be necessary to apply battens to the underside of the rafter in the room area only to obtain sufficient thickness of insulation. Alternatively, it will be necessary to use the higher tech and therefore likely more expensive insulation material to fit between the rafters to gain the required insulation. In a warm roof, however, water storage tanks and piping do not need insulation (unless of course it is a hot water storage tank), but in such structures it is necessary to provide ventilation at the rafter level and at the ridge to ensure moisture is not trapped within the rafter and insulation.

There follow some typical construction details for insulation and ventilation with vapour permeable and vapour impermeable underlays (Figures 8.5–8.7).

CHOOSING THE ROOF COVERING

The choice of roof covering is affected by many things, and consideration of each factor will influence the specification of the tiles, slates or shingles. The major items to be considered are as follows:

Pitch or slope of the roof
Degree of exposure to wind and rain
Weight of the roof covering – what the roof structure is able to carry
Aesthetics
Local planning requirements, if any
Proposed life of the building
Building regulation requirements
Cost
Green issues

Pitch or Slope of Roof

This may be determined by the architecture of the roof, the use which is proposed for the roof space, if any, or indeed the tile type to be used if this has already been specified by planning. Or, if on an extension, the need to match the existing roof covering.

See Figure 8.8. For a quick guide, this illustration can be used to check the roof covering limitations for pitch; these limitations are for normal exposure. If the building is in exposed coastal regions or at high altitude, then the proposed tile or slate manufacturer should be consulted. The minimum pitch indicated for plain tiles and single lap interlocking tiles will generally increase as exposure increases.

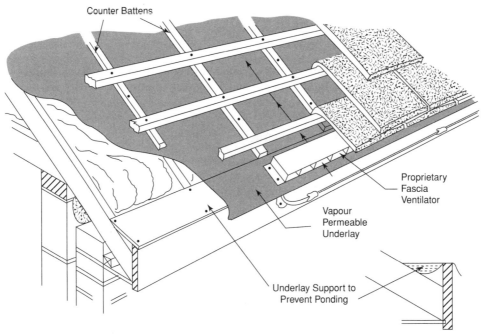

Figure 8.5 Typical construction for a vapour permeable underlay in a cold roof with insulation at ceiling level (unventilated loft).

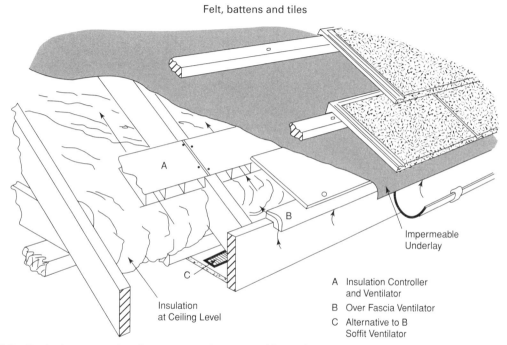

Figure 8.6 Typical construction for a vapour impermeable underlay in a cold roof with insulation at ceiling level.

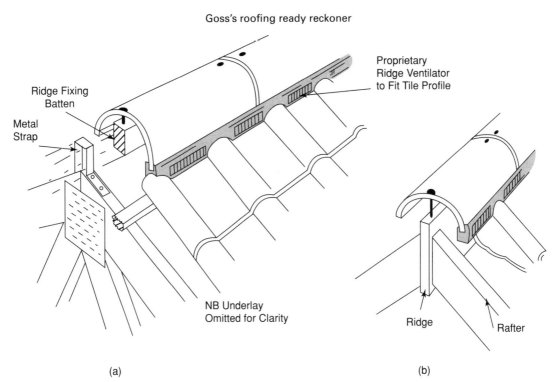

Figure 8.7 (a) Typical ventilated ridge – trussed rafter roof. (b) Alternative ridge fixing – traditional cut roof.

Felt, battens and tiles

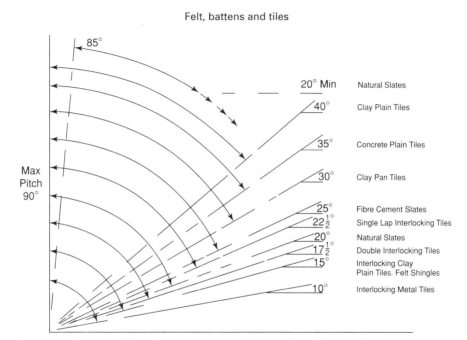

Figure 8.8 Minimum pitch for slates, tiles and shingles for roofs with normal exposure.

Whilst slates may be satisfactory down to 20° pitch in normal or sheltered conditions, they may be limited to a minimum of 40° pitch in severe exposed locations. With plain tiles, slates, and single interlocking concrete tiles, the adjustment can be made by increasing the lap of one tile over the other to improve the tile's performance. Double interlocking tiles are those where there is an interlocking system on both the top of the tile and the side of the tile. They have definite maximum and minimum pitch levels, but because of the interlocking mechanism at the head of the tile, they will generally be satisfactory at a lower pitch than the single side interlocking variety.

Degree of Exposure to Wind and Rain
This matter has been mentioned above, but is worthy of further consideration. What affects exposure?

(a) Degree to which the building is protected by other buildings and or trees
(b) Height of the building above sea level
(c) Proximity to coastline
(d) Location geographically in the UK

If there is a choice of roof covering to be made, then obtain full technical information from a range of tile, slate or shingle manufacturers before starting to specify, and certainly before buying. As has been stated, the lap of the tile or slate will generally be increased with exposure, and this of course means that there will be an increase in number of tiles used on the roof, and therefore the associated cost. By choosing the double interlocking tile, if possible, which may at first appear more expensive, the as-laid cost may in fact be less. The National House-Building Council (NHBC) has guidance on this in their *Standards* document Chapter 7, Appendix 7.2-c, Table 1, which gives guidance on the minimum head lap. For plain tiles, for instance, a 65 mm minimum lap is recommended for normal exposure, but this increases to 75 mm for severe exposure. Further information on the NHBC can be found in the Bibliography.

The height above sea level can be gained from information which was discussed under the roof design chapter in this book. Proximity to coastline is a geographical fact, and there are charts and maps available from the roof tiling manufacturers which indicate the wind exposure and indeed the rainfall exposure throughout the UK. All of this should be taken into consideration.

Loading on the Roof
Two situations must be considered. Firstly, is the building a new building? Secondly, is the new roof covering being fixed to an existing roof?

On new construction, clearly the roof timbers can be designed to accommodate the roof covering chosen, and if a cut roof, information contained in the Tables in Chapter 3, or indeed the TRADA document from which they are taken, can be used. However, for trussed rafter roofs the trussed rafter manufacturer will need to know the type of tile to be used in order to include this in the calculations for the roof structure.

Re-covering an existing roof structure will mean a structural check on the condition of the roof timbers and their fixings. Re-covering with the same type of tile or slate may be safe to do, but the minimum repair would be to replace all the existing battens and renew the underlay (if any was fitted originally), giving due consideration to the specification for the underlay with regard to potential condensation and insulation; information set out in the previous section.

Replacing original clay plain tiles with cheaper concrete plain tiles can save over 5% on the load on the roof, a fact which could be critical with an old roof structure. If there is any doubt about the safety of the roof structure, a structural engineer should be consulted and any suspect timbers replaced or supplementary timbers added to strengthen the structure.

One approach where other factors such as planning constraints do not demand like-for-like replacement, is to assess the safe load that the roof structure can carry by using the services of a structural engineer, and then select a covering within the roof's safe load carrying capacity. A light roof covering such as asphalt (bitumen felt) or cedar timber shingles could be a consideration. The former are available in a range of colours and tile shapes to mimic traditional tiles. The weight of felt shingles which needs an under decking (or sarking) on which they are fixed, will only be approximately 20% of that of the weight of plain tiles. They do produce an impermeable layer, however, and therefore ventilation of the roof void is absolutely essential.

A quick guide to roof covering weights by type is shown in Figure 8.9. The data given are an average; check with the manufacturer for exact loadings, and check the head lap required for the designed exposure, as this also increases the effective load of the tiles on the roof. This factor applies particularly to natural slates.

Aesthetics
How important are aesthetic considerations? Is the roof in an isolated area? What type of roof covering is characteristic of the locality? Is there a planning requirement imposed? Do you have to replace like-for-like on a renovation? Is there a colour preference? Do you wish to match a nearby existing roof? When all of the above applicable questions have been answered, the roof covering will have been specified or at least filtered down to a very limited range.

Local Planning Requirements
Always check with local planning authorities when replacing an existing roof covering, especially if the building is in a conservation area or indeed is itself a listed building. Most new buildings will require planning consent so again, a check with the local planning officer will help make the choice of tile covering, if only by colour. It is too late and can be very costly if the individual's choice of re-roofing material conflicts with that of the planning guidelines, as the planners will have the authority to insist that the roof covering is replaced to their requirements.

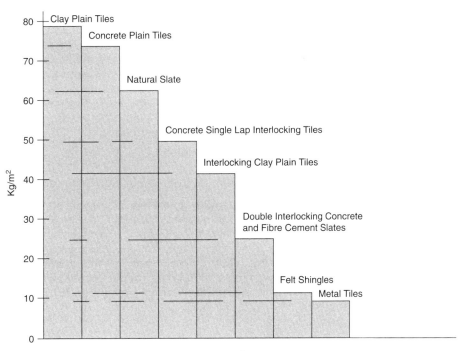

Figure 8.9 Approximate roof covering weights in kg/m².

Proposed Life of the Building

This may seem a strange consideration, but if you have the freedom of choice and the building's use is not long term, then some shorter life roof coverings may be appropriate. Clay tiles for instance will last for 100 years, concrete tiles and slates for 50 years, fibre cement slates for 30 years, and bitumen felt sheeting for 10–15 years depending on quality. All of the above data are of course affected by exposure to sun, wind and rain, and of course the level of maintenance provided.

Building Regulation Requirements

The building regulations state the obvious, i.e. that the roof covering must:

Resist the passage of rain and snow to the inside of the building
Not be damaged by rain or snow
Not transmit moisture due to rain or snow to another part of the building that may be damaged

They also quote the appropriate British Standards codes of practice for precast concrete cladding (non load bearing) and for design and installation of natural stone cladding and lining.

There is also a fire rating requirement where roof coverings pass from one occupancy to another, i.e. semi detached or terraced house situations. Also, where the roof is in close proximity to a boundary. If this situation exists or the project under consideration creates any of the above conditions, consult the local authority building control department and obtain data on the proposed cladding to be used with regard to its performance. Concrete and clay tiles generally present no problems, but check for felt sheet and felt and timber shingles.

Cost

Three basic questions have to be asked before any comparison can be made between various roof coverings available.

Felt, battens and tiles

1. Is the project DIY, i.e. the labour of laying and fixing the tiles is NOT to be part of the cost comparison?
2. Is the project to employ professional roofers on a supply and fix basis?
3. Is it the intention on the project to purchase the tiles direct, but employ a professional roofer to fix them?

In both situations where the building owner is intending to purchase the tiles direct, a careful check should be made on the minimum haulage charges, as these can be quite steep for relatively small quantities.

When considering the cost of the materials themselves (without the cost of fixing) there is surprisingly little difference in cost. The natural materials are generally more expensive than the man made when comparing on the square metre as laid basis. Clearly there are more battens required for small plain tiles than larger interlocking concrete tiles and this too must be taken into account. If the project is DIY, then the speed of laying the tiles may be of consideration, with the larger double interlocking concrete tiles being much faster to cover the roof than plain tiles. This aspect will also of course have an effect on subcontract fixing costs.

As a summary on cost, then, natural slates are likely to be the most expensive roof covering, moving down through man made slates, to concrete plain tiles and then to double interlocking tiles, which will be as little as 30% of the cost of the natural slate product. One then has to balance these costs against life expectancy. The final comparison will vary from project to project, because to carry out a true comparison, the cost per square metre as laid is only one factor. Hips, valleys, verges, ridges, cutting and fitting around dormers, roof windows, chimneys, etc. will also affect the cost of the project, and especially on DIY projects do not overlook the cost of waste.

Green Issues
This aspect of projects is increasingly a major element for the process of specifying roofing, and the purpose of this text is to introduce the reader to the subject, with guidance to further reading.

Government legislation is becoming more stringent as to the effects of producing building materials, the energy they consume, thus giving rise to carbon emissions, in the building of and living in our homes, working in our offices and factories, and their impact on global warming. ISO 14001 sets out guidance for environmental management systems.

The Building Research Establishment (BRE), through its Centre for Sustainable Construction, has a method for assessing the environmental impact of building materials known as the Life Cycle Assessment (LCA). BRE has also developed an Ecopoint system which helps the specifier 'rate' the building material's 'greenness'. BRE also has a method of assessing a whole building, not just the individual materials, known as Eco Homes; this assesses the environmental impact of the building over its full expected life cycle, i.e. it takes into account the durability of the materials and the initial carbon cost of their manufacture. Green Guides are available from BRE; these categorise specifications into A, B or C ratings, A being the best. In roofing terms, for example, concrete roof tiles are typically rated A, whilst man made polymer/resin slates are likely to fall into category C.

For further reading on this subject refer to *The Green Guide to Specification*, by Anderson, Shiers and Sinclair; further details in the Bibliography. Also Marley Roofing's *Green Guide to Roofing*.

Solar roof tiles are available, and becoming increasingly affordable, to produce the ultimate 'green' roof. Whilst solar electric generating panels and solar heated water panels have been available for many years, these are products which are imposed upon the roof with the roof covering passing beneath them. The solar roof tiles mentioned above are specific to each tile manufacturer, fitting precisely into the line of the roof covering, and are therefore much more aesthetically acceptable. They are readily available from such companies as Marley Roofing and can simply be tiled into the roof as the roof covering proceeds. Clearly an electrician is required to complete the wiring to the house mains electricity system, and of course any unused electricity can then be fed back into the National Grid and sold to the occupant's electricity supplier. The Marley system is called Solar Tile; full details of this can be found in the Bibliography.

9 ROOF COVERINGS – BUILDING DETAIL DRAWINGS

NATURAL SLATES
Natural slate is one of the oldest roof covering materials and although much is now imported, slate quarries still exist in Wales and Cornwall, producing traditional riven slates of variable sizes and colours. Slate is an extremely durable material, lasting up to 100 years, and whilst 'slate grey' is a term used to describe the traditional slate colour, they are also available in a range of colours from the traditional grey through blue and green to heather shades. Laid on battens in the normal way, because the slates have no nibs to hold the slate onto the batten, they must be securely fixed with aluminium or copper silicone bronze nails. The holes for the nails are made by the roof layer using a slate holing machine; the position of the holes from top to bottom or 'head to tail' in slating terminology, is decided by the gauge to which they are to be laid. This gauge will depend upon the roof pitch and exposure. The more severe the exposure, the more the slates will be overlapped one on the other and will at some point on the course be three slates thick. Guidance on the lap and gauge of the slates can be obtained from the slate manufacturer.

    Figures 9.1(a) and (b) give and indication of slating techniques. Figure 9.1(a) gives typical construction for eaves and verge on a barge board. Note the special length slate at the eaves and the double batten. This second batten is needed to allow the eaves slate and the full first slate to be nailed independently – it is not possible to nail through the eaves slate from the slate above without damage. Each slate is

Goss's roofing ready reckoner

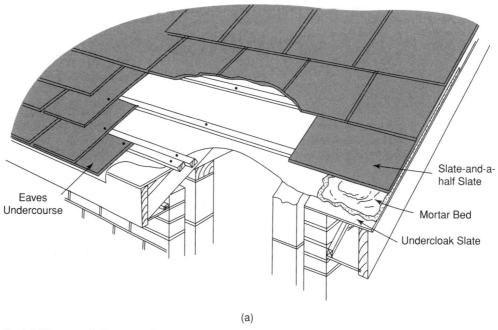

(a)

Figure 9.1 (a) Slate roof. Eaves and verge.

Building detail drawings

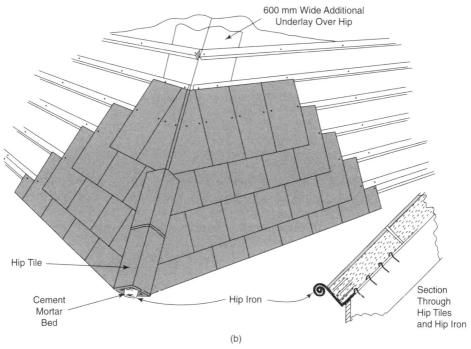

Figure 9.1 (b) Slate roof. Hip detail.

supported on three battens with the batten at the head of the slate far enough above the head to allow the next slate to be nailed directly into the batten. It is for this reason that very careful setting out of lap and gauge of the battens is vital with slate roofing.

Figure 9.1(b) shows one form of slate hip and in addition to the notes above, the key points on the hip are the 'slate-and-a-half' which is shaped by the slate layer to the hip angle. The minimum width at the head of the slate should never be less than 50 mm, as this is its third support point on the batten. Before battens are fixed an additional 600 mm strip of underlay should be laid down the hip length with its centre on the line of the hip board from ridge to eaves. This gives additional water penetration protection and because at the hip line slates can only be butted, the cement mortar on which the hip tile is bedded is the main water proofing for the roof at that point. The hip iron, which traditionally was a forged iron fitting often with ornamental end, is screwed to the hip rafter with galvanised screws; its function to hold the hip tiles in place not only whilst the mortar sets, but later when mortar beds may fail – the iron stops the hip tiles from sliding off the roof. This hip iron is now usually made of stainless steel to prevent corrosion. The end of the hip tile is traditionally filled with pieces of slate bedded in the mortar.

For further information on slate roofing and construction, refer to the slate producer's technical literature. See details in the Bibliography.

CONCRETE INTERLOCKING TILES

Interlocking tiles in their most basic form were made of clay and used in Roman times. These consisted of a tile, semicircular in section, laid alternately like a channel and ridge, i.e. the same tile inverted to form the ridge, thus protecting the joint between the other two. The more common interlocking tiles these days come in two main varieties: firstly, the single interlocking tile where the interlocking shape is on the side of the tile, and the double interlocking tile which has an interlocking shape on both side and head. Single interlocking tiles can be made of clay or concrete, but double interlocking tiles are only made of concrete. They are available in a wide range of colours and profiles, from traditional pantile shape through varying forms of architectural corrugation to imitation slate. Unlike slates and plain tiles, the interlocking tile does not require any overlap of the course below to protect the side butt joints from rain penetration. If you study the slate roof it can be seen that at no point on the roof is there a single layer of slates; there are at least two and at some points, three thicknesses to afford protection. With the interlocking tile, although half bonding, i.e. the tiles laid with staggered joints is traditional, it is not essential for weather proofing. With no interlocking on the head of the tile, the single interlocking tile can have an adjustable head lap to cope with varying exposure conditions and is suitable over a wide range of pitch.

Figures 9.2(a) and (b) illustrate single lap tiles on a barge board verge, eaves and side abutment gutter detail construction. As there is little overlap of the tile, tile clips should be used at the batten as well as nails from tile to batten. It is also vital to use a tile clip at the verge to keep the tiles together and hold them to the roof via the batten. The cement bed at the verge should be seen as a weather proof filler for the joint and not a means of 'sticking' the tiles to the roof.

Most tile manufacturers have a 'dry verge' product, and this consists of a matching colour plastic component which effectively clips the tiles to the verge. Details of such proprietary products can be found from the manufacturer's literature.

Double interlocking tiles have both side and head lap interlocking pattern and are consequently fixed in their course and lap dimension. They do, however, offer an extremely high resistance to wind and water penetration and for that reason can be used at much lower pitches than most other types of tile. As with the single lap tile, clipping is essential, particularly at the verges, as there is very little overlap and therefore very little weight from the tile above on the tile below, thus reducing its wind or indeed suction uplift resistance. Whilst the colour range is similar to that of the single interlocking tile, the interlocking head of the tile tends to limit the profiles to imitation slates and simple architectural corrugations. These tiles tend to be relatively large and because there is minimal lap can be laid very quickly; also because

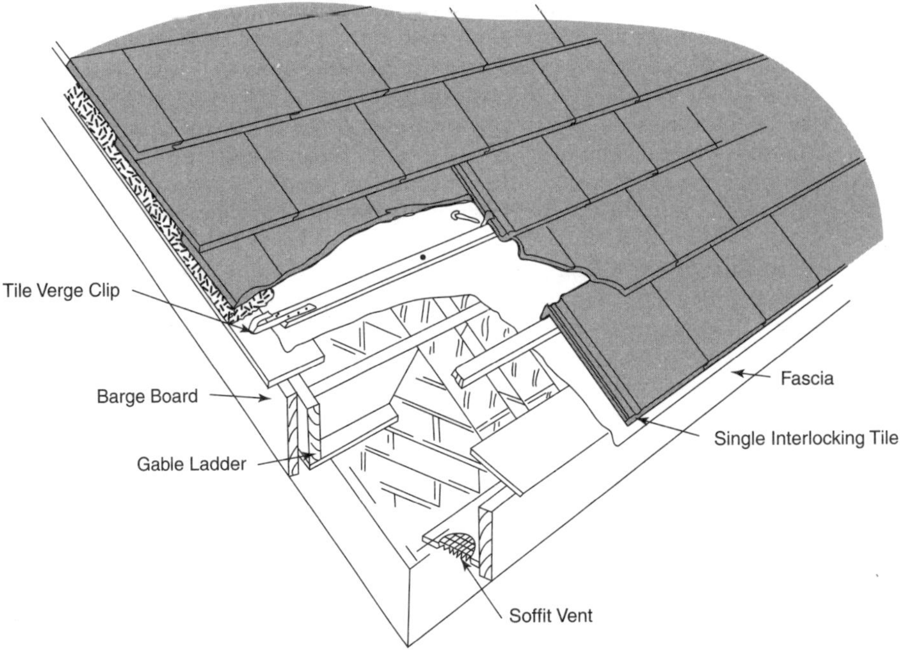

Goss's roofing ready reckoner

Figure 9.2 (a) Single interlocking tile. Barge verge and eaves detail.

156

Building detail drawings

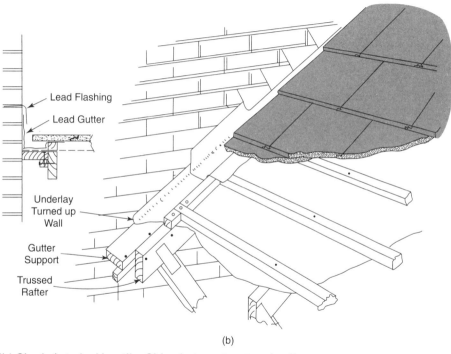

(b)

Figure 9.2 (b) Single interlocking tile. Side abutment gutter detail.

the batten coursing tends to be relatively wide, that process too is fast and economical in the use of batten.

For further information see details of tile manufacturers in the Bibliography.

PLAIN AND PEG TILES
Clay Tiles
Other than thatch, these tiles must be the most traditional form of roofing used in the southern half of England. First used some seven centuries ago, the peg tile made of clay and then fired was a simple rectangle of about 150 mm wide × 250 mm long, pierced towards the top with two square holes through which were driven small wooden pegs. These pegs, often of oak, held the tiles in place over the batten, a function now provided by the clay 'nib'.

Clay plain tiles developed over the years, these being slightly larger at 165 mm wide × 265 mm long and included one or two nibs depending upon the maker and with two or three round holes for nails to securely fix the tiles to the batten. The peg tiles hitherto had simply been laid on the batten with only their weight to keep them in place, with the peg hooked over the batten. Being hand made and with some distortion in firing, the tiles developed to have a slight curve towards the roof in both their width and length, thus ensuring the meeting tiles discharged their water to the tile below. Being made of a natural clay, there was and still is a limited range of colours available, these being influenced by the type of clay in the locality of the tile maker. Colours traditionally range from 'brick' red to darker heather colours. Modern concrete versions of the plain tile are made to the same size and shape as stated above, but with a much wider range of colour choice.

Due to their small individual size, they are more 'flexible' to the roof shape and it must be remembered that seven centuries ago roofs were constructed of selected branches rather than neatly sawn timbers, and consequently were not the flat, even slope of today's trussed rafter roofs. Traditional 'cottage' roofs also necessitated the roofing of small dormers with their associated ridges and valleys and these small tiling units coped extremely well with both the unevenness of the roof and the inaccuracy and irregularity which resulted from the use of irregular timbers for the roof construction. The availability of special sized tiles, i.e. half width tiles, $1^1/_2$ width tiles, special shaped hip tiles and valley tiles, all helped to make this type of roof extremely popular. All of these 'special shape' tiles are still available today.

Laying Plain Tiles

Pitch is critical, because with no side or head interlocking device, like slates, plain tiles rely on overlapping each other both sideways, (minimum 1/3 tile width lap) and head (minimum 65 mm) to provide weatherproofing. With their curved shapes and the irregularity of hand production, it is advisable to use a high quality underlay and to ensure that it is well lapped and dressed at the eaves with good support. At hips, the use of the extra 600 mm width of underlay as indicated in Figure 9.1(b) is essential, and good overlaps at valleys as indicated in Figure 9.3(a) is strongly advisable.

Nailing to battens is even more essential with this relatively light weight small tile to avoid wind displacement. Alloy nails typically 38 mm long × 3.35 mm diameter should be used, with two nails in each tile on the two eaves and ridge courses, and on each of the fifth course up the rafter. At verges and abutments every course should be nailed. As the specified battens for plain tiles are often only 25 mm thick (38 mm wide) for either a 400 mm or 600 mm spacing of the rafter, a high quality timber is paramount because of the high number of nails required on certain battens between each rafter. For this reason it is strongly recommended that 38 mm thick battens be used, especially as battens can often be delivered undersize, but this extra thickness must be taken into account when setting out barge boards and fascias.

Mortar Bedding

Traditionally lime mortar would be used for bedding clay products, and its inclusion in any mix should be used today. Unglazed clay is a breathable product, and can deteriorate relatively quickly if pure cement mortars are used. It is therefore strongly recommended that the mortar for bedding at any point on the roof should be made up of one part of lime to five parts of sharp sand with only one part of cement, which gives the mortar some added weather resistance where it is exposed. For concrete plain tiles, conventional mortar mixes of approximately one part cement to three parts of sand should be used.

Ventilation
Today's manufacturers of both clay and concrete plain tiles offer various forms of in-tile ventilators; some, on a traditional clay tile, manufacture an in-roof ventilator comprised of three separate clay tiles, whilst others integrate plastic ventilation systems. Care must be taken in choosing such ventilators, which should of course be incorporated in any newly constructed roof whether on renovation or new build. The aesthetics of the ventilators may need discussing with your local Conservation Officer.

As always consult the tile maker's literature, some details of whom can be found in the Bibliography under 'Plain and Peg Tiles'.

Plain Tile Valley Detail
See Figure 9.3(a). This is an open valley lined with lead, but in plain tiles there are alternative special valley tiles manufactured which allow the courses to run continuously around the valley. These valley tiles are simply laid in position on the battens and, being V shape, wedge themselves between the tiles of the two abutting roofs. There is also a 'laced' valley, which requires extremely precise setting-out of the tiling battens, allowing the tiles to overlap in a similar manner to the lacing of lines on shoes. For further information on these valleys please refer to manufacturer's information.

Referring to Figure 9.3(a), the tiles at the valley are a cut-tile-and-a-half; this allows a good width of tile at the batten to allow two nails to fix the tile in position. Cutting a single tile could result in only one nail, which would not adequately secure the tile. The valley boards, usually of exterior grade plywood

Building detail drawings

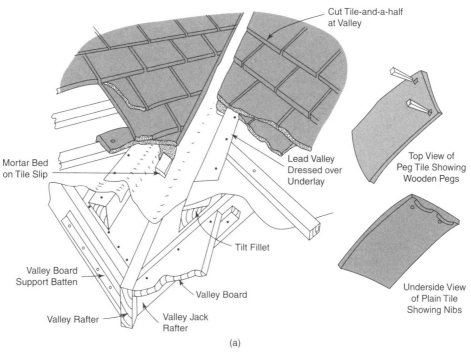

Figure 9.3 (a) Plain tile. Valley detail.

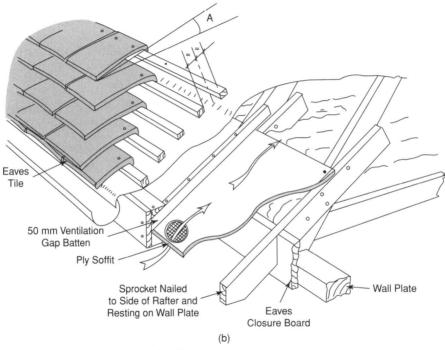

Figure 9.3 (b) Plain tile. Sprocket eaves detail.

or treated timber, should be fitted between the rafters on a support batten and nailed to the side of the rafters. A tilt fillet cut specially to shape, supporting the felt to avoid ponding, carries the cut edge of the tile to maintain the correct pitch. It is usual to dress the lead in one continuous width from tile fillet down and into the valley gutter and up over the next fillet, securing with alloy nails. There should be at least 150 mm lap if the lead has to be joined in the length of the gutter.

A tile slip is then laid in line with the cut edge of the tiles and the tiles bedded on mortar to the mix described above.

This illustration also shows the difference between a peg tile with its wooden pegs in square holes, and the plain tile with its nibs and round holes for nails.

Plain Tile Sprocket Eaves Detail

A simple eaves detail with the fascia fixed directly to the foot of the rafter is illustrated in Figure 9.1(a). Figure 9.3(b) illustrates a sprocket eaves, an architectural feature often found on traditional cottages with a steep pitch, the sprocket helping to arrest the flow of water by lowering the pitch at the eaves. The difference in pitch can be seen at the angle A. In this particular illustration, exposed rafter feet are also illustrated, again a common feature with traditional cottages, which also leads to an exposed soffit that originally would have been left as the underside of the tile and batten. This soffit with its ventilator has a ventilation gap batten fitted on top of it to support the roofing underlay, and maintains the ventilation void. The eaves closure board secures the eaves against insect and bird ingress, and also controls the insulation.

The eaves tile, typically 190 mm long, should project well into the gutter, and careful setting out of the battens is necessary to give equal batten spacing either side of the pitch change, as indicated by the equal signs in the illustration. Care must be taken to ensure that the lower pitch of the sprocketed area is within the minimum recommended by the tile manufacturer.

ASPHALT SHINGLES

Bitumen felt roofing is usually associated with low cost non-dwelling buildings such as sheds, garages, workshops, etc., and the resulting roof covering gives the appearance of a typically green (although other colours are available) flat sheet with little aesthetic appeal. Bitumen felt shingles, however, offer an economic, durable and aesthetic answer to all types of building, where a low weight roof covering is of advantage. These shingles are not the thin felt often associated with the rolls of felt available in builders' merchants but are substantial, up to 3 mm thick, and comprised of a glass fibre mat to give mechanical strength. They are pre-impregnated with bitumen to ensure a solid mat and further bitumen is then added to ensure stability and resistance to temperature fluctuations. Coloured ceramic granules are then bedded in the surface, the range of colours matching closely those of traditional clay or concrete roofing. The underside is coated with silicone sand and a special thermosetting adhesive is then applied to the top surface which bonds the tiles together when laid. The overall effect is that of a tiled roof, with each shingle being divided into three, four, or five tiles, the smaller number mimicking a slate roof, and the larger number representing a plain tiled roof.

Being bonded to the deck of the roof, the shingles can be laid to a very low pitch, typically down to 14°, but even pitches lower than that are possible with special torch-on-underlays applied to the deck before the shingles are laid. It can be seen, therefore, that this is a very versatile product, coping with valleys, hips, ridges and abutments with ease, and can easily be undertaken by a competent DIY enthusiast.

Structure

The shingles require a deck of usually exterior plywood to be laid over the rafters, typically a minimum of 12 mm thick. The verges and eaves should have a metal trim fitted (available from the manufacturers), with each length of trim (typically 3 m long) overlapping by 75 mm and well nailed to the deck. It is advisable to apply the special bitumen mastic adhesive over the nail heads for additional protection and where the sections lap. The eaves course of tiles is cut from a standard shingle and is laid on bitumen mastic applied on the metal eaves trim with the pre-applied shingle mastic pointing down to the eaves. Nails should not be fixed through the pre-applied adhesive but five nails per-shingle should be fitted above the mastic line through the eaves trim and into the roof deck. The nails to be used are galvanised clout nails 8 mm longer than the thickness of the deck. This ensures the nail fully pierces the deck; the fibres grip the nail and resist it springing out. This fact must be considered carefully if it is intended to leave the underside of the deck open to view inside the building, perhaps a garage or workshop. The next course of tiles is then laid conventionally with the decorative tile surface down to the eaves and again nailed through the first eaves course and above the adhesive line. Subsequent courses are laid to half lap, thus imitating conventional tiling.

In cold weather it may be necessary to gently warm the shingles before applying them to the roof, and furthermore the pre-applied adhesive may need warming to perform a satisfactory bond. Failing that, additional mastic gun applied bitumen adhesive (supplied by the manufacturer) can be added to ensure that the tiles are satisfactorily bonded to the course below.

The ridge is formed by cutting the tiles from their shingle, shaping them as indicated in Figure 9.4, bonding and nailing them over the ridge. If the building is located in a particularly windy area, then the lap of the ridge should be downwind to avoid any possibility of the ridge tiles lifting.

Ventilation

By applying the felt shingles to the roof, with or without an underlay, a vapour permeable layer has been created, and ventilation of the roof space is essential. Manufacturers provide in roof ventilators, and ventilation should be provided at eaves and ridge or if the building has no ceiling; then, rather than penetrating the ridge, ventilation can be provided high in the gable ends.

For further information refer to the Bibliography under the heading 'Asphalt Shingles.'

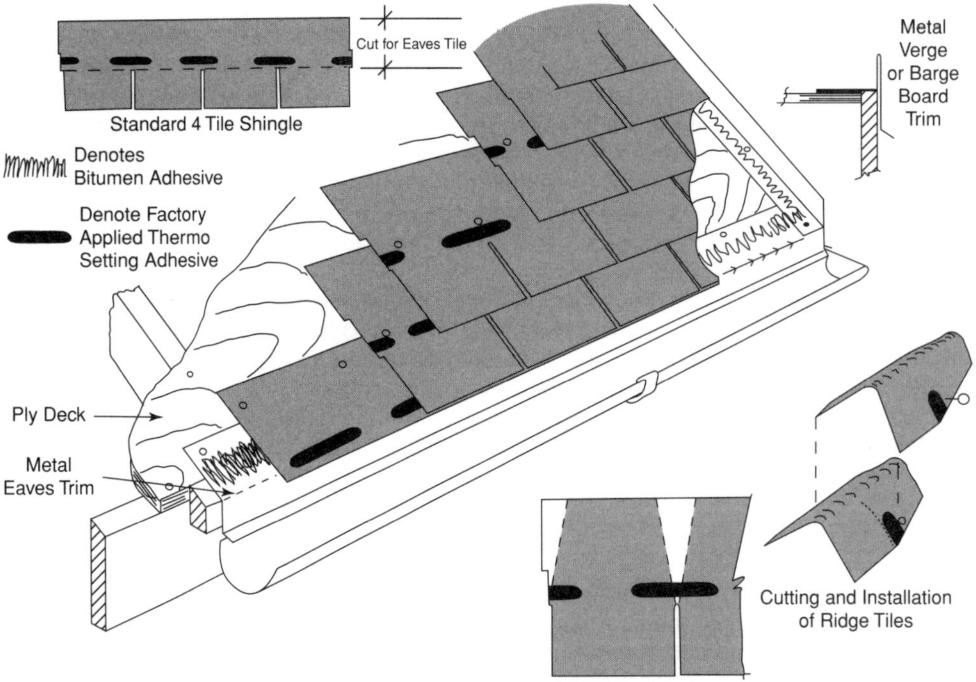

Figure 9.4 Asphalt shingle roofing details.

METAL TILES

Like the asphalt tiles in the previous section, metal tiles are very common in Europe, but gain limited favour in the UK. In a similar way to the asphalt shingle, the metal tiles are available in strips comprising a number of tiles of various corrugated shapes pressed into the metal. The metal is then coated with rust resisting compounds, and the top surface coloured and textured with further applications to give the appearance of traditional tiling. Again, the product offers a very light weight alternative to traditional tiling, can be laid to a minimum pitch of 10°, but unlike the asphalt shingle, does not require a solid deck on which to be laid. Instead they require a conventional rafter roof (or indeed a sarking finished roof) and battens, or in the case of the sarking roof, battens and counter battens. The battens to which the tiles are fixed are generally 50 × 50 mm treated softwood battens, the tiles being fixed with proprietary 50 mm long nails. As some of the nails are left exposed, these need to be coated with the special touch-up paint available from the tile manufacturers.

Unlike other roofing, metal tiles are laid with the first course at the top of the roof, with the head being nailed to the batten, and then the next course down being fitted by lifting the upper course, sliding the tile under and nailing through the lower edge of the upper tile at one of the high points on the corrugation to avoid water ingress. Verges or bargeboard cover profiles, abutment weathering and flashings, and ridges are all provided by the manufacturer as matching pressings. It should be noted, however, that some special equipment is required to both cut the tiles and bend the tiles at hip junctions and abutments. Cutting can be done by tin snips, but this leaves a messy and not very safe edge, or by sheet metal cutters. All cut edges should be treated with the maker's special touch-up products.

Ventilation

Again, being a relatively thin metal roof, and although some air passage is possible though the tile lap, it can only be considered minimal and therefore the roof is 'impermeable'. Adequate ventilation must

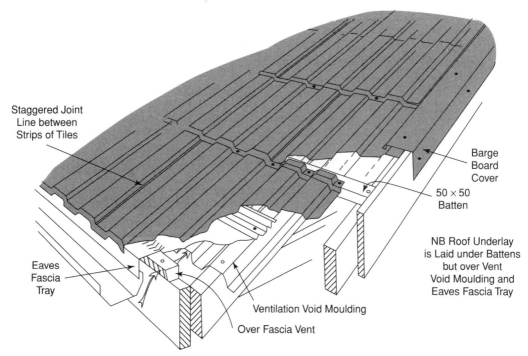

Figure 9.5 Metal tiles – typical details.

be provided and special matching fittings are available from the manufacturers. These include vents for installation at the fascia, within the roof rafter length, and at both the ridge and the hip situation.

Figure 9.5 illustrates typical steel tile roof covering details, and is modelled on Catnic's Lite Tile product. For more details see Bibliography.

10 TOOLS AND EQUIPMENT

The roofing carpenter will need a number of tools and pieces of equipment to satisfactorily obtain information from drawings, mark the timber, set out the roof on the wall plate, cut the timber, and check the completed roof for line, level and plumb. A conventional pencil or pen may be needed for paper calculations, but the true carpenter's pencil should be used for marking timber.

OBTAINING INFORMATION FROM THE DRAWING
(a) Scale rule: only to be used if dimensions are not clearly shown on the drawing.
(b) Protractor: to measure the angle of the roof, but again only if the angle is not written on the drawing.
(c) To mark out the length and angles to be cut on the timbers: steel measuring tape of minimum 5 m in length now available with a digital display to remove the possible error of misreading.
(d) A bevel: a simple carpenter's adjustable bevel is adequate, this being set to the protractor to obtain cutting angles.
(e) Alternatively, a combination square with centre head and built-in protractor is more versatile.
(f) Alternatively, a digital bevel may be used for instant visual display of the setting angles required.
(g) A traditional roofing square can be used if the carpenter is trained in the use of this particular tool.

TO CUT THE ROOF
(h) Hand saw.
(i) Mains or 110 volt electric hand saw. Cordless powered hand saws are available, but a mains power source is still required to recharge batteries.

(j) A compound angle mitre saw. This is an electrically powered saw designed especially to cut angles on timbers – check that the saw is large enough to cope with the length of the cut required; a 300 mm diameter saw should be adequate. This type of saw can tilt in both planes and therefore be set to cut compound angles, i.e. the ridge and edge bevels on hip jack rafters, in one operation. This type of saw is invariably fixed to a bench or stand, and therefore support will be needed for long timbers to be cut. This support should be fitted with a sliding stop system to allow quick repeat lengths to be cut without remeasuring. NB for safety all power tools on site should be 110 volts – for home use a 240 volt saw may be used only in conjunction with a power protection plug adapter – this will protect against accidentally cut cables and faulty wiring, possibly causing electric shock to the operator.

SETTING UP THE ROOF STRUCTURE

(k) A steel tape at least as long as the roof wall plate itself.
(l) A good level, by this we mean a good quality level at least 900 mm long with two spirit levels, one for horizontal use and one for vertical use for plumbing timbers.
(m) Alternatively a good quality level as above but with digital angle readout would be useful to check work on the roof as it proceeds.
(n) To check and set level and verticals over longer distances (i.e. beyond the 900 mm of the levels itself), a laser level could be used. A wide range of laser beam projecting levels are now available giving an accurate beam projection of up to 50 m. The sophisticated rotary laser levels will give both horizontal and vertical beam projection, but it must be remembered that these are accurate only if they are mounted on a stable structure.

ROOF COVERINGS

Whilst many of the pieces of equipment and tools mentioned above will be required in the setting-out, the fitting of the underlay, and the cutting, fitting and nailing of the roof tiling battens, the number of individual and specialist tools required for the various types of roof covering are almost too numerous to mention. Certainly some form of tile cutting device will be required and this may well be of a power saw type, and there will clearly be mortar to mix, with the requirement for some accurate method of measuring the various proportions of sand, lime and cement, etc.

Cutting holes in slates will need a special slate holing tool, and whilst the cutting of asphalt shingles requires no more than tin snips or indeed a good sharp knife, metal tiles should be cut with a metal powered disc cutter, and will require specialist bending equipment available from the manufacturers.

If lead is to be used on the roof, then a range of lead dressing tools will be required, traditionally in the form of wooden mallets and beaters which 'dress' the lead to shape.

Never try to save money by not buying the correct tool for the job. The correct tool, often developed over many centuries by skilled craftsmen, will make the job both easier, faster, and give a more professional result.

11 HEALTH & SAFETY CONSIDERATIONS

ACCESS TO THE ROOF
For anything other than minor maintenance to the roof, a full scaffold or properly designed system of scaffold towers should be constructed in the area on which work is to be carried out. If this is the entire building, then scaffolding will be needed around the entire building. For professional builders, this is likely to be undertaken by a professional scaffolding company, who will visit the site if necessary and advise on the scaffolding required, deliver and erect and adjust them from time to time as may be necessary for safe access to the work. Whilst the professional construction industry is covered by the many rules and regulations including the 'Working at Height' regulations which came into force on the 6 April 2005, the DIY builder and the self builder who is essentially carrying out the work himself and not employing others, is not covered by the regulations. It is still wise to have qualified professionals build the scaffolding and access platforms that may be required, if only for insurance reasons should an accident occur and a claim need to be made on personal insurance.

BASIC PRINCIPLES
Assess the risks involved in each operation.
Plan a safe method of working for each operation.
Decide what personal protection equipment will be required for each operation.

RESTORATION AND RENOVATION OF EXISTING ROOF STRUCTURES

If there is any doubt about the structural soundness of the roof structure itself, this should be carefully examined by a competent person to ensure that it will be safe to work on, and below.

Decide on the access required, taking into account any additional loads on the scaffolding which may occur from the temporary storage of carefully removed roof slates or tiles. If the roof covering material is to be disposed of, ensure that there is a rubbish chute discharging directly into a skip; do not simply throw the tiles to the ground.

Roof structures generally present a problem of access from inside the building, assuming that a full scaffold is to be erected on the outside of the building including the gable ends. It is quite likely, on a restoration project, that the ceilings will be removed even if the ceiling joists are satisfactory and being left in place. Whether the ceiling is lath and plaster or indeed a sheet material, it is best removed from above rather than below because of the fragments, falling elements of the ceiling, and the usual copious amounts of dust. If working from below, then full personal protection equipment should be used, i.e. safety helmet, goggles rather than glasses, the appropriate respiratory mask, stout gloves and steel toe and soled shoes or boots. It may also be worthwhile using one of the disposable all-in-one overalls.

When removing old timber, the rusty nails with which they are connected are a major hazard and should be treated with great respect. Good stout gloves are a necessity for such work, and all nails should be removed where possible or at least bent down and hammered well into the timber for safety. It is likely that replacement timbers will be preservative treated, the preservative itself giving rise to a further Health & Safety hazard which must be addressed. All treated timbers should be delivered dry but recutting and retreating with preservative on site necessitates handling with liquid proof gloves, wearing safety glasses to prevent splashing in the eyes, and a waterproof apron and boots. From an environmental safety viewpoint, the timber treated offcuts should be properly disposed of at a local authority waste disposal centre. They should not be dumped with general rubbish, nor should they be burnt on site.

NEWLY CONSTRUCTED ROOFS

Again, proper scaffolding should be provided all around the perimeter of the roof. If working on a single storey building, or on a building where the floor immediately below the ceiling level which is being worked on has been decked out, then 'soft landing bags' should be used to mitigate the consequences of a fall. Erecting a new roof is a notoriously hazardous part of any building operation, and this applies to traditional roof construction where the support structure is progressively put in place, i.e. ceiling binders, purlins, ridge etc., but particularly to trussed rafter construction where both the hoisting of the prefabricated component, its safe handling whilst being manoeuvred into position, and its temporary fixing in position all pose their own problems. On large scale house building sites, it is now becoming more common for the roof structure to be built on the ground in it entirety, including in some instances felt and battens, and then hoisted into position by crane. This clearly eliminates a number of the hazards in erecting the roof structure.

THE ROOF COVERING

The application of the roof underlay and battens clearly has risks of falling through the roof structure itself, and this can greatly be reduced by laying scaffold boards safely over the ceiling joists of the roof being worked on. When working on an underlay and battened roof, great care must be taken to walk up and down the roof, only on the line of the rafter.

All roof coverings are heavy, some considerably more than others. Whilst steel tiles are relatively light for the area they cover, they are likely to be loaded onto the roof in packs of several kilos each. With plain tiles on the other hand, whilst individually not heavy, the sheer quantity required for the roof covering will add almost 1 ton to an average sized house or bungalow. Consideration should be given to exactly how these are going to be lifted to the roof, and if they are going to be temporarily stacked on the scaffolding, which will have to be designed to carry their weight. Great care must be taken to ensure that the roof is loaded equally on both sides of the ridge when loading-out the roof with its tiles or slates. Care must also be taken not to place large stacks of tiles on the roof, but to spread the stacks evenly immediately above a rafter and equally up the rafter length, thus mimicking the load which the rafter will eventually take. Temporary overloading of the rafter can cause buckling if it is not adequately restrained by diagonal bracing, and because the buckling can be passed from one rafter to another via the tile battens, distortion at the gable end can occur.

CONCLUSION

The above is by no means intended as an exhaustive list for the safe construction of roofs and their coverings. The number of operations involved from the bedding of the wall plate to the placing of the last tile or slate on the roof involves:

Sawing of timber by hand saw or power saw
Use of preservative treatments
Application of nails by hammering
Drilling with hand or power drills for bolts
Lifting of sometimes heavy steel or timber components
Power cutting of clay, concrete and metal tiles
Use of sharp knives and tin snips for asphalt shingles
Constant handling risks whilst laying grit covered concrete tiles
Sharp edges of freshly cut slates and tiles
Use of cements and mixing of mortar for bedding the roof covering
Finally, the possible use of plasticisers and accelerators in the mortar.

Carefully consider all aspects of Health & Safety during construction of the project; a serious accident could mean that you will not complete it.

Safe building!

BIBLIOGRAPHY

DESIGN & CONSTRUCTION

Mindham, C. (2006). *Roof Construction & Loft Conversion*, 4th edn. 256 pp. Blackwell Publishing, Oxford. ISBN 1-4051-3963-3. Available through bookshops.

National House-Building Council (NHBC), Buildmark House, Chiltern Ave, Amersham, Bucks HP6 5AP
Tel. 0870 241 4302
www.nhbc.co.uk

Span Tables for Solid Timber Members in Floors, Ceilings and Roofs (Excluding Trussed Rafter Roofs) for Dwellings. Technical publication published by TRADA Technology Design Aid, DA1/2004, ISBN 1900510464. Available from TRADA Technology, Chiltern House, Stocking Lane, Hughenden Valley, High Wycombe, Bucks, HP14 4ND.
Tel. 01494 569600
www.trada.co.uk

ENVIRONMENTAL ISSUES

Anderson, J., Shiers D. & Sinclair M. (2002). *The Green Guide to Specification*, 3rd edn. 112 pp. Blackwell Publishing, Oxford. ISBN 0-632-05961-3. Available through bookshops.

Green Guide to Roofing by Marley Roofing. Available from Marley Roofing, Station Road, Coles Hill, Birmingham, B46 1HP.

Solar Roofing Systems by Marley Roofing. Available from Marley Roofing, Station Road, Coles Hill, Birmingham, B46 1HP.
www.marleyroofing.co.uk/solar

Marley Technical Advisory Service: Tel. 08705 626900

INSULATION

Thinsulex Multi Layer
Progressive Products Ltd, Industrial Estate, Presteigne, Powys, LD8 2UF.
  Tel. 01544 260500

Therma Fleece – Sheep's Wool Insulation
Second Nature Ltd, Soulands Gate, Dacre, Penrith, Cumbria, CA11 0JF.
  Tel. 01768 486285
  www.secondnatureuk.com

Rockwool
Rockwool Ltd, Pencoed, Bridgend, CF35 6NY.
  Tel. 01656 862621
  www.rockwool.co.uk

Froth-Pak – Polyurethane Foam
Dow Chemical Co Ltd, Station Road, Sandbach, Cheshire, CW11 3JG.
  Tel. 0800 3694 637

Renotherm – Polyurethane Foam Application Service
ISC Renotherm Ltd, New Street House, Petworth, West Sussex, GU28 0AS.
  Tel. 01798 345401
  www.iscrenotherm.co.uk

Warmcel – Fibre Information Application Service
Excel Industries, Maerdy Industrial Estate South, Rymney, Gwent, NP22 5PY.
  Tel. 01685 845200
  www.excelfibre.com

Thermapitch and Kooltherm – Rigid Foam Insulation
Kingspan Insulation, Pembridge, Leominster, Herefordshire, HR6 9LA.
  Tel. 01544 388601
  www.insulation.kingspan.com

METAL CONNECTORS FOR TIMBER ROOFS
Simpson Strong-Tie, Winchester Road, Cardinal Point, Tamworth, Staffordshire, B78 3HG.
  Tel. 01827 255600
  www.strongtie.co.uk

Fasten Master. UK distributor: OSC, 63a, Victoria Road, Burgess Hill, West Sussex, RH15 9LN.
  Tel. 0845 2419862
  Technical information – Timber Solve Ltd: 01420 549201

TILES, SLATES & SHINGLES
Asphalt Shingles
Matthew Hebden, 54 Blacker Moor Road, Sheffield, S17 3GJ.
  Tel. 0114 236 8122
  www.matthewhebden.co.uk

Clay Plain Tiles, Concrete Tiles, Slates
Sandtoft – contact through general enquiries: 0870 1452020
  www.sandtoft.co.uk
Eternit Building Materials, Ridgehill Drive, Madeley Heath, Crewe, Cheshire, CW3 9LY.
  Tel. 01782 758800
  www.eternit.co.uk

Concrete Tiles
Russell Roof Tiles Ltd, Wellington Road, Burton on Trent, Staffordshire, DE14 2AW.
  Tel. 01283 517070

Fibre Cement Slates
Marley Eternit Ltd, Station Road, Coles Hill, Birmingham, B46 1HP.
  Tel. 01283 722103
  www.eternit.co.uk

Natural Slates
Welsh Slate, Penrhyn Quarry, Bethesda, Bangor, Gwynedd, LL57 4YG.
  Tel. 01248 600656
  www.welshslate.com

Plain and Peg Tiles
Keymer Tiles Ltd, Nye Road, Burgess Hill, West Sussex, RH15 0LZ.
  Tel. 01444 232931
  www.keymer.co.uk

Eternit Building Materials, Ridgehill Drive, Madeley Heath, Crewe, Cheshire, CW3 9LY.
 Email: marketing@eternit.co.uk
 Tel. 01782 758800

Plain Tiles, Single and Double Interlocking Tiles, Interlocking Slates

Marley Building Materials, Station Road, Coles Hill, Birmingham, B46 1HP.
 www.marleyroofing.co.uk
 Technical information: 08705 626900

Sandtoft – contact through general enquiries: 0870 1452020
 www.sandtoft.co.uk

Steel Tiles

Catnic, Pontypandy Industrial Estate, Caerphilly, CF83 3GL.
 Tel. 01292 0337900
 www.corusconstruction.com

Index

A

Asphalt shingles, 164–6

B

Barge boards, 156, 168
Battens, 124–9
Bevel cut, 28, 30, 34
Binder, 12
Birds mouth, 6, 7
Bolts, 122
Bracing,
  diagonal, 113
  longitudinal, 6, 7, 113
  wind, 112–13

C

Clay tiles, 158
Cold roof, 124, 134
Collar, 23
Concrete tiles, 154–8
Condensation, 132

Counter batten, 126, 140
Cutting angles, 21–109

D

Design
  roof timber, 8–20
Diagonal bracing, 113
Dormers, 112–16
Dry ridge, 142
Dry verge, 155

E

Eaves details, 140–41, 152, 162,
    166, 168
Edge bevel, 30

F

Fascia, 6, 7
Fascia vent, 140
Felt, 124
Framing anchors, 122

G

Gauge
  tile and slate, 151–4
Green issues, 149

H

Health & safety, 173–6
Hip
  iron, 153
  rafter, 6, 7, 23
  tile, 153

I

Imposed load, 15
Insulation, 129–32
Interlocking tiles, 154–8

J

Joist
  ceiling, 4, 5

L

Lap
  slate, 154
  tiles, 155
  underlay, 127

Lead
  dressing tools, 172
  flashing, 157
  gutter, 157, 161
Level
  laser, 171

M

Metal connectors, 120–23
Mortar bedding, 159

N

Nails, 122, 128–9

O

Openings in roofs, 112–19

P

Peg tiles, 158–63
Pitch, 2, 3, 24, 139
Pitch limit chart, 143
Plain tiles, 158–63
Ponding, 140
Preservative treatment, 128, 174, 176
Purlin, 2, 3, 23, 35

R

Rafter
  common, 2, 3, 23
  jack, 2, 3, 23, 29
  valley Jack, 4, 22, 29, 33
Rafter hip, 23, 30, 31
Rafter trussed, 113, 117
Ridge bevel cut, 28, 30
Ridge cut, 27, 33
Ridge dry, 142
Ridge ventilator, 142
Rise, 21, 25, 27
Roof abutment, 157
Roof openings, 112–19
Roof windows, 118–19
Roofing network, 120

S

Sarking, 126, 138
Saws
  compound mitre, 171
Screws, 122
Seat cut, 27
Shingles
  asphalt, 164–6
Slates, 151–4

Snow loads, 9, 16, 18
Soffit, 6–7
Soffit vent, 141, 156, 162
Solar tiles, 150
Span, 21–5
Sprocket, 162–3
Steel tiles, 167–9
Strapping
  gable, 110–11, 120
  wall plates, 110, 120

T

Tile
  clip, 156
  cutting, 161, 172
  weights, 147
Tiles
  peg, 158–63
  plain, 158–63
  interlocking, 154–8
  solar, 150
  steel, 167–9
Timber strength classes, 9
Trimmers, 114–16
Truss clip, 120–22
Trussed rafter, 4, 5, 113, 117

U

'U' value, 129
Under
  cloak, 152
  lay, 124, 140, 141

V

Valley, 5, 22, 29, 161
Valley board, 161
Vapour, 124, 133
Ventilation, 129
Ventilation void, 132–8, 162, 168
Ventilator
  fascia, 140
  ridge, 142

W

Wall plate, 2, 3, 23, 110
Warm roof, 124, 133–7
Weights of tiles and slates, 147
Wind bracing, 4, 5, 112
Wind load, 9